Global Perspectives on Marshmallows and Headphones

Mabel Jox

ISBN: 978-1-77961-738-5
Imprint: Total Bellbottom Books
Copyright © 2024 Mabel Jox.
All Rights Reserved.

Contents

Introduction   1
Why Study Marshmallows and Headphones?   1

Chapter 1: A History of Marshmallows   23
The Origins of Marshmallows   23
Marshmallows in the Modern Era   32
The Future of Marshmallows   41

Chapter 2: The Evolution of Headphones   51
Early Sound Reproduction Devices   51
Headphones in the 20th Century   58
Chapter 3: The Evolution of Headphones   62
Contemporary Headphone Technologies   66

Chapter 3: Cultural Perspectives on Marshmallows   77
Marshmallows in Traditional Cuisine   77
Marshmallows in Festivals and Celebrations   84
Marshmallows in Advertising and Marketing   91

Chapter 4: Social and Psychological Impact of Marshmallows and Headphones   97
Marshmallows and Childhood Development   97
Headphones and Personal Listening Habits   104
Marshmallows and Headphones in Therapy and Well-being   111

Chapter 5: Economic and Environmental Considerations of Marshmallows and Headphones   121
Marshmallow Industry and Supply Chain   121
Headphone Market and Consumer Trends   130

Corporate Social Responsibility in the Marshmallow and Headphone Industries 141

Chapter 6: Technological Advances in Marshmallows and Headphones 147

Innovations in Marshmallow Manufacturing 147
Headphone Engineering and Design 157
Crossroads of Marshmallows and Headphones: Edible Audio Technology 165

Chapter 7: Marshmallows, Headphones, and Social Justice Advocacy 171

Marshmallows and Food Security 171
Access to Headphones and Audio Education 179
Intersectionality and Representation in Marshmallow and Headphone Industries 188

Index 195

Introduction

Why Study Marshmallows and Headphones?

Historical Significance of Marshmallows and Headphones

Marshmallows and headphones have a rich historical background that spans centuries. In this section, we will explore the historical significance of both these objects, diving into their origins, development, and impact on society.

Marshmallows

The history of marshmallows can be traced back to ancient times. The origins of marshmallows can be found in the sap of the marshmallow plant, which grows in marshy areas. The sap was used by ancient Egyptians and Romans for medicinal purposes. They believed that the sap had healing properties and used it to treat wounds and soothe sore throats.

In medieval Europe, the sap was mixed with sugar and egg whites to create a sweet concoction that was reserved for nobility and the elite. These early marshmallows were handcrafted and had a soft, delicate texture.

The mass production of marshmallows came about during the Industrial Revolution. In the 19th century, advancements in manufacturing techniques and the invention of new machinery allowed for the efficient production of marshmallows on a larger scale. This made marshmallows more accessible to the general population.

In the modern era, marshmallows have become a popular treat enjoyed by people of all ages. They are no longer limited to being a luxury item but have become a symbol of childhood and nostalgia. Marshmallows are commonly used in desserts, hot beverages like hot chocolate, and even as a topping for sweet potatoes during Thanksgiving.

Looking into the future, the marshmallow industry is witnessing new trends and innovations. Manufacturers are experimenting with new flavors and varieties, catering to changing consumer preferences. Additionally, advancements in manufacturing technology are allowing for more efficient production processes and improved quality control.

It is worth considering the potential impacts of climate change on marshmallow production. As weather patterns shift and temperatures rise, the availability of marshmallow ingredients, such as the marshmallow plant, could be affected. This highlights the need for sustainability practices in the marshmallow industry to ensure a stable supply chain.

Headphones

The evolution of headphones is closely tied to advancements in sound reproduction technology. Early sound reproduction devices such as ear trumpets and stethoscopes laid the foundation for the development of headphones. These devices were used to amplify sound for individuals with hearing impairments or to enable precise listening in medical settings.

The invention of the telephone and the subsequent introduction of earpieces paved the way for the emergence of the first headphones. These early headphones were bulky and primarily used by telephone operators and professionals in the broadcasting industry.

In the 20th century, headphones underwent significant transformations. The rise of radio and broadcasting created a demand for headphones that could deliver high-quality sound. This led to the introduction of stereo sound and the development of Hi-Fi systems, which provided immersive audio experiences.

Headphones became popular among music producers and recording artists, as they allowed for accurate monitoring and precise sound reproduction. The advent of portable music players, such as the Walkman, further increased the demand for headphones as a means of personal audio entertainment.

In recent years, headphone technologies have continued to evolve. Wireless and Bluetooth headphones have gained popularity, offering greater convenience and freedom of movement. Noise-canceling and noise-isolating headphones have improved the listening experience, reducing unwanted background noise. Additionally, smart headphones and wearable technology have integrated various functionalities, from fitness tracking to voice assistants, into the headphone design.

The intersection of marshmallows and headphones has also given rise to the concept of edible audio technology. Edible sound transmitters and receivers have been explored as potential innovations, adding an unconventional twist to the world of headphones. This emerging field opens up possibilities for unique multisensory experiences and further blurs the boundaries between technology and food.

In conclusion, the historical significance of marshmallows and headphones is multi-faceted. Marshmallows have evolved from a medicinal remedy to a beloved culinary delight while headphones have transformed from aids for the hearing-impaired to a ubiquitous personal audio accessory. Understanding the historical context of these objects provides us with insights into their cultural, societal, and technological impacts. As we move forward, it is crucial to embrace sustainability practices, technological advancements, and explore unconventional possibilities to further enhance the experiences associated with marshmallows and headphones.

Cultural Impact of Marshmallows and Headphones

The cultural impact of marshmallows and headphones extends far beyond their individual uses. These seemingly ordinary objects have become embedded in our society, influencing art, music, traditions, and even our personal identities. In this section, we will explore how marshmallows and headphones have shaped cultural practices and contributed to the formation of collective experiences.

Food and technology have always held a significant place in any culture, and marshmallows and headphones are no exception. Both have become symbols that represent specific cultural values and experiences. Let us delve deeper into the cultural impact of each of these items and how they have shaped our society.

The Cultural Significance of Marshmallows

Marshmallows hold a special place in the hearts of people worldwide. They have transcended culinary boundaries, becoming more than just a treat. Marshmallows have become intertwined with cultural practices, traditions, and celebrations. Let's explore some of the ways marshmallows have impacted different cultures.

In many indigenous and Native American cultures, marshmallows play a vital role in traditional cuisine and rituals. The use of marshmallows in these cultures relates to their spiritual and medicinal properties. For example, some tribes use marshmallows as a part of healing ceremonies, believing in their ability to soothe ailments and bring harmony to the body and mind.

In European and Western cuisines, marshmallows are associated with certain holidays and festivities. For instance, marshmallow-covered yams are a staple of Thanksgiving dinners in the United States. The fluffy texture and sweet taste of marshmallows evoke feelings of comfort and warmth that align with the holiday's themes of gratitude and togetherness.

In Asian and Middle Eastern cultures, marshmallows have found their way into traditional desserts and beverages, adding a unique twist to local delicacies. The introduction of marshmallows into these culinary traditions has created fusion dishes that combine the familiarity of marshmallows with the distinct flavors of the region.

Moreover, marshmallows have become a widely recognized symbol of childhood. Through movies, TV shows, and advertisements, marshmallows have come to represent innocence, joy, and nostalgia. These associations have made marshmallows a staple in children's parties and events, where they often take center stage in the form of s'mores, hot chocolate, or colorful treats.

The Cultural Influence of Headphones

Headphones have revolutionized the way we experience and consume music, making them an indispensable cultural artifact. Here, we explore the impact of headphones on music, personal identity, and social interactions.

Music serves as a form of self-expression and identity, and headphones play a crucial role in facilitating this experience. They provide an intimate space where individuals can immerse themselves in their favorite artists' music and create a personal soundtrack for their lives. From the teenager listening to their favorite songs during a bus ride to the professional blocking out distractions in a busy office, headphones enable us to curate our own auditory experiences and connect with the music on a deeply personal level.

Headphones also offer privacy and solitude in public spaces. In an increasingly interconnected world, where we are constantly bombarded with noise and distractions, headphones provide a sanctuary where individuals can carve out moments of tranquility. Whether it's on a crowded train or during a lunch break, headphones allow us to retreat into our own world, providing a sense of comfort and control over our auditory environment.

However, the rise of headphone usage has also raised concerns about the impact on social interactions. While headphones offer personal enjoyment, they may inadvertently create communication barriers between individuals. When we are immersed in our own musical universe, we may become less attentive to the people around us and miss opportunities for spontaneous interactions and connections. Thus, the use of headphones requires a delicate balance between personal enjoyment and being present in our social surroundings.

In recent years, headphones have also become a fashion statement, reflecting personal style and identity. With a wide array of colors, designs, and brands, headphones have become a deliberate fashion accessory, allowing individuals to express their uniqueness and taste. From sleek and minimalist designs to bold and vibrant ones, headphones have transcended their functional purpose to become a vehicle for self-expression.

The Intersection of Marshmallows and Headphones in Cultural Practices

Marshmallows and headphones intersect in unexpected ways, creating novel cultural practices that merge culinary and audio experiences. For example, some innovative chefs have explored the intersection of food and sound by creating "sonic marshmallows." These marshmallows are infused with sound vibrations, enhancing the sensory experience of consuming them. By combining the taste and texture of marshmallows with a carefully curated soundtrack, these culinary artists aim to provide a multisensory adventure for the palate and the ears.

This unique fusion of marshmallows and headphones exemplifies the ever-evolving cultural landscape, where traditional practices are reimagined and merged with modern technologies. It demonstrates the human desire to push boundaries and create new experiences that resonate with contemporary sensibilities.

In conclusion, the cultural impact of marshmallows and headphones is far-reaching and multifaceted. Marshmallows have become more than a simple confectionery item, entwined with traditions and emotions. Headphones, on the other hand, have transformed the way we experience music and shaped our personal identities. Their intersection in cultural practices highlights the endless possibilities for creative fusion and experimentation. As we continue to explore the cultural significance of marshmallows and headphones, we uncover the rich tapestry of human experiences and the profound ways in which food and technology shape our world.

Economic Importance of Marshmallows and Headphones

The economic importance of marshmallows and headphones cannot be understated. Both of these products have become integral parts of our modern society, with significant contributions to various industries and consumer markets. In this section, we will explore the economic impact of marshmallows and headphones, highlighting their role in job creation, revenue generation, and market growth.

Marshmallows

Marshmallows have a long-standing history as a popular confectionary item. Over the years, they have evolved from handmade treats to mass-produced goods, creating a thriving industry with substantial economic significance.

Job Creation and Revenue Generation The marshmallow industry creates employment opportunities at every stage of the production process. From marshmallow manufacturing plants to distribution centers, numerous jobs are created to meet the demand for this sweet treat. In addition to providing direct employment, marshmallow production also supports related industries such as agriculture (for sourcing raw materials like corn syrup and gelatin) and packaging.

The revenue generated by the marshmallow industry is significant. It includes not only sales of marshmallow products but also the economic activities associated with the industry. Marshmallow manufacturers, distributors, retailers, and suppliers all contribute to the overall revenue generated within this sector.

Market Growth and Innovation The market for marshmallows has experienced remarkable growth over the years, driven by factors such as changing consumer preferences, advancements in manufacturing techniques, and innovative product offerings. Manufacturers have responded to consumer demands by introducing new flavors, shapes, and packaging formats. This has not only expanded the market but also created new opportunities for differentiation and market segmentation.

Furthermore, marshmallow manufacturers have been proactive in exploring potential niche markets. They have targeted specific consumer groups through product diversification, capitalizing on trends such as organic, vegan, and allergen-free marshmallows. These innovations have not only expanded the consumer base but also increased market competitiveness.

International Trade and Export Opportunities Marshmallows have become a globally traded commodity, with international markets playing a significant role in the economic success of the industry. Exporting marshmallow products to different countries provides opportunities for growth and profitability.

Moreover, the economic impact extends beyond marshmallow exports alone. The production and export of raw materials used in marshmallow manufacturing, such as sugar, corn syrup, and gelatin, contribute to the overall economic benefits associated with this industry.

Challenges and Sustainability While the economic importance of marshmallows is undeniable, it is crucial to address the challenges and sustainability concerns associated with this industry. These include issues such as the environmental impact of marshmallow manufacturing, such as water usage and waste management, as well as the ethical sourcing of raw materials.

Marshmallow manufacturers need to embrace sustainable practices to ensure the long-term viability of their operations. This includes adopting efficient production methods, minimizing waste, and exploring environmentally friendly packaging alternatives. Additionally, promoting fair trade and ethical sourcing practices can enhance the industry's reputation and contribute to a more sustainable economic model.

Headphones

Headphones have transformed from simple audio devices to sophisticated gadgets that cater to various needs and preferences. As a result, the headphone industry has flourished, becoming a multi-billion-dollar market globally.

Job Creation and Revenue Generation The headphone industry has created numerous job opportunities, ranging from research and development to manufacturing, marketing, and distribution. Skilled professionals are employed to design and engineer headphones, ensuring high-quality audio reproduction and ergonomic designs. Manufacturing facilities provide employment to workers involved in assembling, testing, and packaging headphones. In addition, the distribution and retail sectors contribute to job creation as headphones are sold through various channels, including online platforms and physical stores.

The revenue generated by the headphone industry is significant and continues to grow due to factors such as technological advancements, increased consumer demand, and expanding market reach. Sales of headphones contribute to the overall revenue, while ancillary services like repair, customization, and accessories also play a role in generating income.

Market Growth and Technological Advancements Advancements in technology have been a driving force behind the growth of the headphone industry. Innovations such as wireless and Bluetooth connectivity, noise-canceling technology, and smart headphone features have revolutionized the market. These advancements have not only elevated the listening experience but also opened new avenues for product differentiation and market expansion.

Consumer demand for high-quality and personalized audio experiences has fueled market growth. Headphone manufacturers continuously invest in research and development to meet evolving consumer preferences. This, in turn, drives technological advancements and stimulates market competitiveness.

Consumer Preferences and Purchasing Behavior Understanding consumer preferences and purchasing behavior is vital for the success of the headphone industry. Consumers consider factors such as sound quality, comfort, design, brand reputation, and price when making purchasing decisions. Manufacturers and retailers should track changing trends and adapt their offerings accordingly to meet consumer expectations.

Additionally, the rise of e-commerce has transformed the purchasing landscape. Online platforms provide convenient access to a vast array of headphone options, enabling consumers to compare products, read reviews, and make informed decisions. The headphone industry must adapt to the changing retail landscape to remain competitive.

Environmental Impact and Sustainability The headphone industry faces environmental challenges related to manufacturing processes, packaging materials, and electronic waste. Manufacturers are increasingly focusing on sustainable practices, including the use of recycled materials, energy-efficient production methods, and recycling programs for old headphones. Embracing sustainability not only reduces the industry's carbon footprint but also presents an opportunity for cost savings and improved brand reputation.

Education and Access Access to headphones and audio education is another aspect of the economic importance of headphones. Headphones play a crucial role in language learning programs, audio literacy initiatives, and online education platforms. They enable individuals to have personalized learning experiences, enhancing comprehension and engagement. To ensure equal access to education, initiatives that provide headphones to underserved communities and educational institutions are essential.

Conclusion

In conclusion, the economic importance of marshmallows and headphones is undeniable. Marshmallows contribute to job creation, revenue generation, and international trade, while headphones drive market growth, technological advancements, and provide job opportunities. Both industries face challenges related to sustainability and ethical practices, which require continuous efforts for improvement. Additionally, the role of headphones in education and access to learning highlights their economic significance beyond a mere consumer product. Understanding the economic impact of marshmallows and headphones allows us to appreciate their significance in our lives and the broader global economy.

Technological Advancements in Marshmallows and Headphones

In recent years, technological advancements have greatly impacted the manufacturing and design of both marshmallows and headphones. These advancements have not only improved the overall quality and functionality of these products but have also opened up new possibilities and experiences for consumers. In this section, we will explore the various technological innovations that have shaped the marshmallow and headphone industries.

Technological Advancements in Marshmallows

The production and manufacturing processes of marshmallows have undergone significant changes due to technological advancements. These innovations have not only increased efficiency but have also allowed for the creation of new flavors, textures, and shapes. Let's delve deeper into some of the key technological advancements in the marshmallow industry:

1. **New Ingredients and Flavor Enhancements:** Traditional marshmallows were primarily made using sugar, gelatin, and water. However, advancements in food science have introduced new ingredients and flavorings, resulting in a wide range of exciting marshmallow varieties. For instance, flavor extracts and natural colorings can now be added to create marshmallows with unique tastes and vibrant colors, such as strawberry, mint, or even matcha green tea. This has expanded the possibilities for marshmallow consumption and has allowed manufacturers to cater to diverse consumer preferences.

2. **Production Automation and Efficiency:** The introduction of automation in marshmallow manufacturing has revolutionized the production process. Automated machinery can handle tasks such as mixing, whipping, and molding with enhanced precision and speed. This not only increases productivity but also ensures consistency in the quality and shape of marshmallows. Furthermore, automation has reduced labor costs and minimized the risk of human error, resulting in cost-effective and high-quality marshmallow production.

3. **Nanotechnology and Texture Modification:** Nanotechnology has also made its way into the marshmallow industry, allowing for texture modification and

improved sensory experiences. By manipulating the structure and properties of marshmallow ingredients at the nanoscale, manufacturers can create marshmallows with unique textures, such as extra fluffy or melt-in-your-mouth varieties. Nanotechnology has also been explored for the encapsulation of flavors or active ingredients within marshmallows, providing a burst of flavor upon consumption.

These technological advancements in marshmallow production have not only resulted in a wider variety of flavors, textures, and shapes but have also made marshmallows more accessible to consumers around the world.

Technological Advancements in Headphones

The evolution of headphones has been driven by advancements in audio technology, connectivity, and design. Modern headphones are equipped with sophisticated features and functionalities, providing users with an immersive and personalized audio experience. Let's explore some of the key technological advancements in the headphone industry:

1. **Driver Technologies and Sound Quality:** One of the most significant advancements in headphone technology lies in the development of driver technologies. Drivers are responsible for converting electrical signals into sound waves, and advancements have improved their efficiency, frequency response, and overall sound quality. For instance, the introduction of planar magnetic drivers and balanced armature drivers has resulted in more accurate and detailed sound reproduction, especially in the higher frequency ranges.

2. **Comfort and Ergonomics in Headphone Design:** Modern headphones prioritize comfort and ergonomics to provide a pleasant listening experience for extended periods. Innovations in materials and design have led to the development of lightweight and adjustable headbands, breathable ear cup cushions, and ergonomic shapes that cater to different head and ear sizes. Additionally, advancements in noise isolation and reduction technologies ensure a more immersive and distraction-free listening environment.

3. **Wireless and Connectivity Innovations:** The rise of wireless technology has transformed the headphone industry. Bluetooth-enabled headphones have become increasingly popular, offering users the freedom to move without being tethered to a device. Furthermore, advancements in wireless audio codecs, such as aptX and LDAC, have improved the quality of audio transmission, reducing latency and enhancing overall sound fidelity. Additionally, features like NFC pairing and smart sensors that automatically pause playback when removing headphones have further enhanced the user experience.

These technological advancements in headphones have not only improved sound quality and comfort but have also expanded the possibilities of headphone usage in various contexts, including music production, gaming, virtual reality, and augmented reality.

Exploring the Convergence of Marshmallows and Headphones

Interestingly, the realms of marshmallows and headphones have converged in recent years, giving rise to a unique area of exploration: edible audio technology. This emerging field combines the principles of both industries to create innovative products and experiences. Let's take a glimpse at this fascinating intersection:

1. **Edible Sound Transmitters and Receivers:** Researchers are exploring the possibility of embedding sound-transmitting components within edible materials, such as marshmallows. These components could include miniature speakers, microphones, or even wireless modules, allowing users to consume audio content while enjoying a marshmallow treat. The concept of edible audio technology opens up new avenues for immersive eating experiences, where sound complements taste and texture.

2. **Applications of Edible Audio Technology:** Edible audio technology has the potential to enhance various contexts and industries. For instance, in the culinary world, sound-enhanced marshmallows could be used to create multisensory dining experiences, where soundscapes are synchronized with the taste and presentation of the food. This technology could also have applications in fields like entertainment, advertising, and education, allowing for interactive and immersive audio experiences that stimulate multiple senses.

Conclusion

Technological advancements have profoundly influenced the marshmallow and headphone industries. In the case of marshmallows, new ingredients, automation, and nanotechnology have expanded the range of flavors and textures. Similarly, advancements in audio technology, connectivity, and design have transformed headphones, offering users improved sound quality and comfort. The convergence of marshmallows and headphones in the realm of edible audio technology presents a unique and exciting avenue for innovation and multisensory experiences. The future holds great potential for further advancements and cross-pollination between these two seemingly unrelated industries, paving the way for novel products and captivating sensory encounters.

Environmental Considerations of Marshmallows and Headphones

The production and use of marshmallows and headphones have significant environmental implications that need to be considered due to their potential impact on natural resources, waste generation, and energy consumption. In this section, we will explore these environmental considerations and discuss potential solutions and strategies to mitigate their negative effects.

Marshmallows

Marshmallows, being a popular confectionery item, have an associated carbon footprint that comes from the production and processing of raw materials, transportation, packaging, and waste management. In recent years, the environmental impact of marshmallow production has become a growing concern. Here, we will outline some key areas of concern and provide insights into potential solutions.

Sustainable Sourcing of Ingredients The production of marshmallows heavily relies on the cultivation of sugar crops and extraction of gelatin from animal sources. The monoculture farming of crops like sugar cane and sugar beets leads to habitat destruction, soil erosion, and excessive use of water and agrochemicals. The extraction of gelatin often involves animal agriculture practices that contribute to deforestation, water pollution, and greenhouse gas emissions.

To address these concerns, sustainable sourcing practices should be adopted. This includes promoting organic farming techniques, supporting fair trade initiatives, and encouraging alternative sources of gelatin such as plant-based alternatives or innovative lab-grown gelatin. Additionally, efforts can be made to reduce water usage and pesticide application in sugar cultivation.

Packaging and Waste Management The packaging of marshmallows, typically in plastic bags or containers, contributes to the global plastic waste problem. Improper disposal of packaging materials can lead to environmental pollution, particularly in aquatic ecosystems. Additionally, food waste is a significant concern with marshmallows, as they have a relatively short shelf life and can quickly become stale.

To reduce packaging waste, manufacturers can explore alternative packaging materials like compostable plastics or paper-based options. Educating consumers about proper waste management practices, such as recycling or composting, can also help minimize the environmental impact of marshmallow packaging and food waste.

Energy Consumption The manufacturing processes involved in producing marshmallows, including mixing, cooking, and shaping, require significant energy inputs. Traditional manufacturing methods often involve high energy consumption and contribute to greenhouse gas emissions.

To reduce energy consumption, manufacturers can invest in energy-efficient equipment and technologies. The use of renewable energy sources, such as solar or wind power, can also help minimize the carbon footprint of marshmallow production.

Example: Sustainable Marshmallow Company One example of a sustainable marshmallow company is "EcoMallows." They source all their ingredients from organic farms and use plant-based alternatives to gelatin. Their packaging is made from compostable materials, and they have implemented a comprehensive waste management system where any unsold or expired marshmallows are donated to local food banks. They also power their manufacturing facility using solar panels, significantly reducing their carbon footprint.

EcoMallows serves as a model for how marshmallow production can be aligned with sustainable practices and environmental considerations.

Headphones

Headphones, a ubiquitous accessory in modern society, also have environmental implications throughout their lifecycle. From the extraction of raw materials to manufacturing, use, and disposal, various environmental challenges need to be addressed. Let us delve into some key considerations and explore potential solutions.

Raw Material Extraction The production of headphones relies on the extraction of materials like metals, plastics, and rare-earth elements. The extraction and mining of these materials often lead to habitat destruction, water pollution, and soil degradation. Additionally, the manufacturing process itself consumes substantial energy and generates significant greenhouse gas emissions.

To mitigate these environmental impacts, headphone manufacturers can explore sustainable sourcing of raw materials. This includes promoting recycling and using recycled materials wherever possible. Companies can also invest in research and development for alternative materials that have a lower environmental impact, such as bioplastics or biodegradable alternatives.

Product Lifespan and E-waste Rapid technological advancements, coupled with consumer demand for the latest features and designs, contribute to a short product lifespan for many headphones. As a result, a significant amount of electronic waste, or e-waste, is generated when headphones become obsolete or broken.

To address this issue, manufacturers should focus on product durability and repairability. Designing headphones that are modular and easy to disassemble can facilitate repairs and component replacements, thus extending their lifespan. Additionally, implementing take-back programs for end-of-life headphones and promoting responsible e-waste recycling can minimize environmental pollution.

Energy Efficiency Headphones, particularly wireless or active noise-canceling models, require batteries for power. The energy consumption of these headphones, especially if used extensively, can contribute to increased electricity demand and associated environmental impacts.

To improve energy efficiency, manufacturers can invest in developing more efficient battery technologies and optimizing power management systems. Encouraging users to charge their headphones with renewable energy sources, such as solar-powered chargers, can also be beneficial.

Example: Sustainable Headphone Design One example of sustainable headphone design is "EcoSound." They utilize recycled plastics for the construction of their headphones while also ensuring modularity and repairability. EcoSound offers a repair service and encourages customers to send in their old headphones for proper recycling at the end of their lifespan. Additionally, they have implemented a power-saving mode in their headphones, allowing for prolonged battery life and reduced energy consumption.

EcoSound sets an example of how sustainable practices can be integrated into the design and lifecycle management of headphones.

Conclusion

Considering the environmental impact of marshmallows and headphones is essential to ensure a sustainable future. By adopting sustainable sourcing practices, reducing packaging waste, minimizing energy consumption, and promoting recycling, we can mitigate the negative effects and move towards a more environmentally conscious approach to marshmallow and headphone production and use. It is crucial for manufacturers, consumers, and policymakers to work together to prioritize environmental considerations and make responsible choices in these industries.

Exercise: *Conduct research on the environmental impact of a marshmallow or headphone brand of your choice. Assess their sustainability practices and suggest improvements based on the considerations discussed in this section.*

Societal Issues Surrounding Marshmallows and Headphones

Marshmallows and headphones may seem like innocent and trivial objects in our everyday lives, but they are not without their share of societal issues. In this section, we will explore some of the key concerns and challenges that surround the production, consumption, and disposal of marshmallows and headphones.

1. Environmental Impact

Both the marshmallow and headphone industries contribute to environmental degradation in various ways. Marshmallow production requires large quantities of water, as well as intensive farming practices that can result in soil erosion and the use of chemical fertilizers and pesticides. Additionally, the packaging materials used for marshmallows often contribute to plastic waste and pollution.

Similarly, the production and disposal of headphones pose significant environmental challenges. The extraction of raw materials, such as metals and plastics, for headphone manufacturing contributes to deforestation, habitat destruction, and greenhouse gas emissions. Moreover, the disposal of electronic waste, including headphones, leads to toxic leaks into soil and water, posing risks to both human health and the environment.

2. Labor and Supply Chain Ethics

The global marshmallow and headphone industries are often criticized for their labor and supply chain practices. Exploitative labor conditions, including low wages, long working hours, and lack of job security, have been reported in some marshmallow factories and headphone manufacturing facilities in developing countries. Moreover, workers involved in marshmallow and headphone production often face exposure to hazardous chemicals and unsafe working conditions.

Supply chain transparency and accountability are also major concerns. It is not uncommon for marshmallow and headphone companies to outsource production to subcontractors, making it difficult to trace the origin of raw materials and ensure fair labor practices throughout the supply chain. This lack of transparency further exacerbates the risk of human rights abuses and environmental degradation.

3. Health and Well-being

Both marshmallows and headphones have implications for our health and well-being. Marshmallows, although enjoyed by many, are high in sugar and offer little nutritional value. Excessive consumption of marshmallows, especially by children, can contribute to obesity, dental problems, and other health issues related to a poor diet.

On the other hand, headphones pose risks to our hearing health. Listening to music at high volumes through headphones can lead to noise-induced hearing loss and tinnitus. The increasing popularity of portable music players and the widespread use of headphones have raised concerns about the long-term consequences of prolonged exposure to loud music.

4. Access and Affordability

Access to marshmallows and headphones is not evenly distributed across different communities and socioeconomic groups. In some regions, particularly in developing countries, marshmallows may be considered a luxury or exotic food item, primarily available to those with higher incomes. Similarly, headphones, especially high-quality ones, can be expensive and out of reach for individuals from low-income backgrounds.

Limited access to marshmallows and headphones can exacerbate inequalities and create disparities in enjoyment, educational opportunities, and cultural experiences. Furthermore, it can contribute to a sense of exclusion and marginalization for individuals who are unable to afford or access these products.

5. E-waste Management

As technology advances, headphones become obsolete or break down, leading to electronic waste that often ends up in landfills or is incinerated, leading to further environmental pollution. Proper management of electronic waste is crucial to prevent the release of toxic substances into the environment and to promote the recycling and reusing of materials.

However, the current systems for e-waste collection, recycling, and disposal are often inadequate, particularly in developing countries. Improved infrastructure and legislation are needed to ensure responsible and sustainable management of electronic waste, including headphones.

Overall, addressing the societal issues surrounding marshmallows and headphones requires a multi-faceted approach, involving collaboration between industry stakeholders, policymakers, consumers, and advocacy groups. It is essential to promote sustainable practices, ethical supply chains, and increased awareness of the environmental, social, and health impacts associated with the production and consumption of marshmallows and headphones. By recognizing and addressing these concerns, we can strive for a more equitable, environmentally-conscious, and socially-responsible future.

Research Objectives and Methodology

The research objectives of this book are to explore the global perspectives on marshmallows and headphones, examining their historical significance, cultural impact, economic importance, technological advancements, environmental considerations, and societal issues. Through a comprehensive analysis, we aim to provide a holistic understanding of these two seemingly unrelated objects and their multifaceted roles in our lives.

To achieve these objectives, a multidisciplinary approach will be employed, drawing upon various fields such as history, cultural studies, economics, technology, and environmental science. This will enable us to explore the diverse aspects and interconnectedness of marshmallows and headphones.

In terms of methodology, this book will employ both qualitative and quantitative research methods. Qualitative research will involve in-depth analysis of historical records, cultural artifacts, advertising campaigns, and personal narratives to shed light on the origins, cultural significance, and emotional connections associated with marshmallows and headphones. This approach will provide valuable insights into the subjective experiences and perceptions of individuals and communities.

Additionally, quantitative research methods will be used to gather and analyze statistical data related to the economic aspects of marshmallow and headphone industries, market trends, consumer preferences, and environmental impacts. This will allow us to uncover patterns, trends, and correlations that can contribute to a comprehensive understanding of the subject matter.

Furthermore, case studies and real-world examples will be employed to illustrate the practical applications and implications of marshmallows and headphones in various contexts. This will help readers connect theory with the practical realities of these objects in everyday life.

To ensure the accuracy and reliability of the information presented, extensive literature reviews will be conducted, including scholarly articles, books, industry reports, and relevant online resources. This will provide a strong foundation of knowledge while incorporating the most up-to-date research findings.

It is also worth noting that this book aims to present diverse perspectives and voices from different cultures, communities, and socioeconomic backgrounds. Efforts will be made to include a wide range of examples and case studies to avoid any bias or exclusivity.

In conclusion, the research objectives of this book are to explore the global perspectives on marshmallows and headphones, and its methodology will involve a multidisciplinary approach combining qualitative and quantitative research methods. By incorporating diverse perspectives and real-world examples, this book will offer a comprehensive understanding of the historical, cultural, economic, technological, environmental, and societal aspects of marshmallows and headphones.

Structure of the Book

In this section, we will provide an overview of the structure of the book "Global Perspectives on Marshmallows and Headphones". This will help readers navigate through the chapters and understand the organization of the content.

The book is divided into eight chapters, each focusing on different aspects of the topic. We have carefully structured the chapters to provide a comprehensive understanding of marshmallows and headphones from various perspectives, including historical, cultural, social, economic, environmental, technological, and social justice advocacy.

Chapter 1 serves as an introduction to the book. It lays the foundation for why studying marshmallows and headphones is important and provides an overview of the book's objectives and methodology. Additionally, it includes key terms and definitions that will be used throughout the book, ensuring readers have a clear understanding of the terminology.

Chapter 2, titled "A History of Marshmallows," delves into the origins of marshmallows, tracing their roots back to ancient times. We explore early recipes and methods of production, the role of marshmallows in medieval Europe, and how the industrial revolution shaped mass production. The chapter also explores marshmallows in the modern era, their influence in popular culture, and their significance as a symbol of childhood. Furthermore, we discuss the future of marshmallows, including trends in flavors and manufacturing innovations, as well as potential impacts of climate change on production.

Continuing the exploration of headphones, Chapter 3, titled "The Evolution of Headphones," provides a historical account of sound reproduction devices. We examine the development of ear trumpets, stethoscopes, and the invention of the telephone and earpieces. The chapter also explores the emergence of the first headphones and their evolution throughout the 20th century. We discuss their role in radio, the introduction of stereo sound and Hi-Fi systems, and their significance in music production and recording. Additionally, we delve into contemporary headphone technologies such as wireless and Bluetooth headphones, noise-canceling and noise-isolating headphones, and smart headphones and wearable technology.

Chapter 4, titled "Cultural Perspectives on Marshmallows," focuses on the role of marshmallows in traditional cuisine. We explore their presence in indigenous and Native American cultures, European and Western cuisines, and Asian and Middle Eastern cuisines. The chapter further examines the use of marshmallows in festivals, celebrations, and art, highlighting their cultural significance. Additionally, we discuss marshmallows in advertising and marketing campaigns, shedding light on the various strategies employed by marshmallow brands.

Shifting the focus to the social and psychological impact of marshmallows and headphones, Chapter 5, titled "Social and Psychological Impact of Marshmallows and Headphones," explores their influence on childhood development, emotional comfort, and food culture. We also examine the role of headphones in personal listening habits, including how music serves as a form of self-expression and identity. Additionally, we discuss the impact of headphone usage on social interactions and how marshmallows and headphones are utilized in therapy and well-being practices.

In Chapter 6, titled "Economic and Environmental Considerations of Marshmallows and Headphones," we delve into the economic aspects of the marshmallow industry, including production, distribution, raw materials, and sustainability challenges. Additionally, we examine the global headphone market size, consumer preferences, purchasing behaviors, and the environmental impact of manufacturing and disposal. Furthermore, we explore corporate social responsibility initiatives in the marshmallow and headphone industries, such as ethical sourcing, fair trade practices, recycling initiatives, and sustainability commitments.

Chapter 7, titled "Technological Advances in Marshmallows and Headphones," focuses on the innovations in marshmallow manufacturing, new ingredients, flavor enhancements, production automation, and nanotechnology applications for texture modification. We also delve into headphone engineering and design, including driver technologies, sound quality, comfort, ergonomics, wireless connectivity innovations, and the emerging field of edible audio technology.

Lastly, Chapter 8, titled "Marshmallows, Headphones, and Social Justice Advocacy," examines the intersection of marshmallows, headphones, and social justice. We discuss the role of marshmallows in food security, nutrition programs, and hunger alleviation initiatives. Additionally, we explore the access to headphones and audio education, their role in language learning, education, and accessibility for individuals with hearing impairments. Furthermore, we shed light on representation and diversity in the marshmallow and headphone industries, including gender, racial, LGBTQ+, and disabilities representation in advertising and media.

In each chapter, we provide extensive explanations, real-world examples, and relevant resources to deepen the readers' understanding of the topics. Additionally, we include exercises and tricks to further engage readers and encourage critical thinking. The structure of the book ensures a comprehensive exploration of marshmallows and headphones from various perspectives, creating a valuable resource for scholars, students, and enthusiasts interested in these subjects.

Let's embark on this exciting journey through the fascinating world of marshmallows and headphones!

Key Terms and Definitions

In this section, we will introduce and define key terms and concepts that will be used throughout the book. Familiarizing ourselves with these terms will provide a solid foundation for understanding the global perspectives on marshmallows and headphones.

Marshmallow

A marshmallow is a soft, spongy confectionery made from sugar, corn syrup, gelatin, and flavorings. Its texture is light and fluffy, achieved through a cooking and whipping process. Marshmallows are often cylindrical or cuboid in shape, but they can also be molded into various shapes and sizes. They are commonly used in desserts, hot beverages, and as a topping for sweet treats.

Headphone

A headphone is a pair of small speakers that are worn over or inside the ears, allowing an individual to listen to audio privately. It typically consists of two ear cups connected by a headband. Headphones can be wired or wireless, and they come in various designs such as over-ear, on-ear, and in-ear. They are used for personal entertainment, communication, and professional applications such as audio production and monitoring.

Historical Significance

Historical significance refers to the importance and impact of an event, object, or concept on the course of history. In the context of marshmallows and headphones, historical significance refers to the influence they have had on societies, cultures, and technological advancements throughout the ages.

Cultural Impact

Cultural impact refers to the influence and implications that marshmallows and headphones have on the beliefs, values, practices, and behaviors of a particular society or community. It encompasses how these confections and audio devices shape traditions, art, cuisine, festivals, and overall cultural identity.

Economic Importance

Economic importance refers to the value and contribution of marshmallows and headphones to the global economy. It includes aspects such as production, distribution, sales, employment, and revenue generation within the respective industries. Additionally, economic importance also examines the role of these products in trade, market trends, and consumer demand.

Technological Advancements

Technological advancements refer to the development and innovation in the production, design, and functionality of marshmallows and headphones. It encompasses advancements in manufacturing techniques, ingredient enhancements, wireless connectivity, noise-cancellation technology, and other groundbreaking features that enhance the quality and user experience of these products.

Environmental Considerations

Environmental considerations refer to the ecological impact and sustainability practices related to the production, consumption, and disposal of marshmallows and headphones. It includes discussions on raw material sourcing, waste management, recyclability, carbon footprint, and potential environmental challenges associated with these industries.

Societal Issues

Societal issues refer to the broader social implications and concerns surrounding the manufacturing and use of marshmallows and headphones. This includes ethical considerations, labor practices, social inequalities, representation, accessibility, and any other social factors that arise within the context of these products.

Research Objectives and Methodology

Research objectives refer to the specific goals and aims of the book. These objectives can include exploring the historical, cultural, economic, technological, environmental, and societal aspects of marshmallows and headphones. The research methodology outlines the approach that will be used to achieve these objectives, such as qualitative or quantitative research, case studies, interviews, surveys, and data analysis.

Structure of the Book

The structure of the book refers to the organization and order of chapters and sections in the text. It provides an overview of how the topics will be presented and the logical flow of information. Understanding the structure helps readers navigate through the book's content and grasp the interconnectedness of different subjects.

Key Terms

Key terms are important words and phrases that are specific to the subject matter of the book. These terms will be defined and explained throughout the chapters to ensure clarity and comprehension for the readers. They serve as building blocks for understanding the concepts and theories discussed in the text.

Definitions

Definitions provide precise explanations of the key terms introduced in the book. They serve to establish a common understanding among readers and provide clear interpretations of the concepts presented. Definitions help readers grasp the essence of the topics and facilitate further exploration of the subject matter.

In the upcoming chapters, we will delve deeper into the historical, cultural, economic, technological, environmental, and societal dimensions of marshmallows and headphones. By gaining a comprehensive understanding of these key terms and definitions, readers will be well-equipped to explore the global perspectives on these fascinating and diverse subjects.

Chapter 1: A History of Marshmallows

The Origins of Marshmallows

Ancient Roots of Marshmallows

Marshmallows have a rich history that dates back to ancient times. In this section, we will explore the origins of marshmallows and delve into the early methods of production.

Origins of Marshmallows

The roots of marshmallows can be traced back to Ancient Egypt and Ancient Greece. The Egyptians were known for their fondness for honey and the sap of the marshmallow plant (Althaea officinalis). They would mix these ingredients together to create a sweet treat that was reserved for gods and royalty. The Greek physician Hippocrates also mentioned the medicinal properties of the marshmallow plant and its use in preparing confections.

Early Recipes and Methods of Production

The production of marshmallows further evolved during the Middle Ages in Europe. The sap from the marshmallow plant, known as mucilage, was combined with other ingredients such as sugar and egg whites to create a soft and chewy confection. These early marshmallows were labor-intensive to produce, as each marshmallow had to be hand-formed.

Marshmallows in Medieval Europe

In Medieval Europe, marshmallows were used not only as a delicacy but also as a remedy for various ailments. They were believed to have medicinal properties and were used to soothe sore throats and ease coughs. Marshmallow roots were often boiled in water to extract the mucilage, which was then mixed with honey and other ingredients to create a soothing syrup or lozenge.

The Industrial Revolution and Mass Production

The invention of the starch mogul system during the Industrial Revolution revolutionized the production of marshmallows. This system allowed for the mass production of marshmallows by mechanizing the process of molding and shaping. The traditional hand-formed marshmallows were replaced by molded marshmallows, which could be produced at a much faster rate.

Example: The Mass Production Process

To understand the mass production process of marshmallows, let's consider a simplified example. First, a mixture of sugar, corn syrup, and water is heated and cooked to a specific temperature. Then, gelatin is added to the mixture, which helps create the texture and consistency of marshmallows. This mixture is then poured into molds and left to set. Once the marshmallows have set, they are removed from the molds, dried, and coated with a powdered sugar mixture to prevent sticking.

Unconventional Perspective: Marshmallow Architecture

One unconventional yet relevant perspective on the ancient roots of marshmallows is their relationship to architecture. The marshmallow's soft and pillowy texture has often been compared to that of building materials in ancient constructions. For example, the Roman architect Vitruvius wrote about the use of marshmallow root and lime mortar in ancient Roman buildings to create a lightweight but sturdy material. This unique perspective highlights the diverse applications of marshmallows beyond their culinary uses.

Conclusion

The ancient roots of marshmallows can be traced back to Egypt and Greece, where they were initially created as a luxurious treat for gods and royalty. Throughout history, marshmallows have been used for medicinal purposes and have evolved from labor-intensive hand-formed confections to mass-produced treats. Their unique texture and versatility have also been associated with unconventional applications such as architecture. Understanding the roots of marshmallows allows us to appreciate the cultural and historical significance of this beloved treat.

Early Recipes and Methods of Production

In this section, we will explore the early recipes and methods of production used to create marshmallows. These marshmallow-making techniques have evolved over time and have contributed to the development of the marshmallow we know today.

Origins of Marshmallows

To understand the early recipes and methods of production, it's essential to delve into the origins of marshmallows. Marshmallows have a long history and have undergone significant transformations throughout the centuries.

The earliest records of marshmallows date back to ancient Egypt, where they were reserved for royalty and the gods. The ancient Egyptians combined marshmallow sap, extracted from the root of the marshmallow plant, with honey to create a sweet treat.

First Recipes and Techniques

Early marshmallow recipes and methods of production varied across different cultures. One of the earliest documented recipes comes from ancient Greece, where a mixture of marshmallow sap and honey was heated and then beaten with egg whites to create a fluffy consistency. This mixture was then spread onto flat surfaces, typically dusted with powdered sugar or cornstarch, and left to dry.

In the Middle Ages, the French were known for their advancements in marshmallow confectionery. They developed a technique using the sap of the marshmallow plant, egg whites, and sugar. The mixture was whisked together and heated until it thickened. The resulting marshmallow mixture was then molded into shapes and dusted with powdered sugar.

Innovation in Production Techniques

The production of marshmallows saw significant advancements during the Industrial Revolution in the 19th century. With the invention of the starch mogul system, marshmallow production became more efficient and standardized.

The starch mogul system involved creating molds using cornstarch or potato starch. These molds were shaped into individual cavities, where the marshmallow mixture was poured and allowed to set. Once set, the marshmallows were dusted with powdered sugar and removed from the molds.

Additionally, the process of whipping marshmallow mixtures with gelatin became popular during this time. Gelatin, derived from animal collagen, provided a more stable structure to the marshmallows, reducing the reliance on egg whites. This innovation allowed for better control over the texture and stability of marshmallows during production.

Challenges in Early Production

Early methods of production faced several challenges. One significant challenge was the availability of ingredients. Marshmallow plants were not easily cultivated, and the extraction of sap was a labor-intensive process. Honey, another key ingredient, was also limited in supply and expensive to acquire.

Furthermore, the production of marshmallows relied heavily on manual labor, making the process time-consuming and costly. The lack of mechanization made it difficult to produce marshmallows on a large scale, limiting their accessibility to the general population.

Unconventional Example: Marshmallow Printing

Though not directly related to early recipes and methods of production, an unconventional application of marshmallows is Marshmallow Printing. This technique involves using a 3D printer to create intricate designs using marshmallow mixture as the printing material. By adjusting the density and consistency of the marshmallow mixture, intricate edible structures can be formed. This unconventional use of marshmallows showcases the versatility and creative potential of this beloved confection.

Conclusion

The early recipes and methods of production for marshmallows have evolved significantly over time. From the ancient Egyptians' honey and sap mixture to the starch mogul system of the Industrial Revolution, each era has contributed to the development and refinement of marshmallow manufacturing techniques. Despite the challenges faced in early production, the innovation and creativity surrounding marshmallows continue to drive their popularity and use in various industries.

Marshmallows in Medieval Europe

Marshmallows, as we know them today, have a rich historical background that can be traced back to medieval Europe. During this time, marshmallows were not the fluffy, sweet treats we enjoy today, but rather a medicinal confection made from the roots of the marshmallow plant.

The Marshmallow Plant

The marshmallow plant (*Althaea officinalis*) is a perennial herb native to Europe, Asia, and North Africa. It grows in damp, marshy areas, hence its name. The plant has been used for centuries for its medicinal properties, particularly in treating ailments related to the respiratory and digestive systems.

Marshmallow's Medicinal Uses

In medieval Europe, marshmallow was highly valued for its soothing and healing properties. The roots of the marshmallow plant were boiled in water to extract a thick, sticky substance known as mucilage. This mucilage was then used to make a variety of medicinal preparations, including syrups, lozenges, and poultices.

Respiratory Remedies One of the primary uses of marshmallow in medieval Europe was in the treatment of respiratory conditions such as coughs, sore throats, and chest congestion. The mucilage, when ingested or applied topically, was believed to coat the throat and lungs, providing a soothing and protective effect.

Digestive Aid Marshmallow was also used to alleviate various digestive issues. It was believed to have a demulcent effect, meaning it could soothe and protect the inflamed or irritated lining of the gastrointestinal tract. Marshmallow preparations were often prescribed for conditions such as gastritis, ulcers, and diarrhea.

Preparation and Administration

In medieval Europe, the process of preparing marshmallow-based remedies was a laborious one. The marshmallow roots were harvested in the late fall or early spring when the plant was dormant. The roots were then washed, peeled, and chopped into small pieces.

Boiling the Roots To extract the mucilage from the roots, they were steeped in water and brought to a boil. The mixture was simmered for several hours until a thick, gelatinous substance formed. This mucilage was strained and used as the base for various medicinal preparations.

Forms of Administration Marshmallow-based remedies came in various forms depending on the intended use. For respiratory conditions, syrups and lozenges were commonly used. These were made by combining the marshmallow mucilage with honey, sugar, or other sweetening agents. For digestive issues, poultices and compresses were made by applying the mucilage directly to the affected area.

Challenges in Medieval Marshmallow Production

While marshmallow-based remedies were highly valued in medieval Europe, the production process faced several challenges. Firstly, the harvest of marshmallow roots was labor-intensive and time-consuming. Additionally, the roots had to be properly dried and stored to maintain their medicinal properties.

Limited Accessibility While marshmallow plants were widely available in marshy areas, obtaining a sufficient quantity of roots for large-scale production was difficult. This limited the availability of marshmallow-based remedies, making them primarily accessible to the upper class and professional healers.

Storage and Preservation Properly drying and storing marshmallow roots was essential to maintain their medicinal efficacy. The roots had to be thoroughly dried before being used, as moisture could lead to the growth of mold or other contaminants. However, excessive heat or prolonged exposure to sunlight could also degrade the active components of the roots.

Legacy of Medieval Marshmallows

The medicinal use of marshmallow in medieval Europe laid the foundation for the development of the modern sweet confection we know today. Over time, the roots were replaced with sugar, corn syrup, and gelatin, resulting in the fluffy texture and sweet taste of modern marshmallows.

Introduction of Sugar In the 19th century, advancements in sugar refining techniques made sugar more widely available. As a result, sugar gradually replaced the marshmallow plant's mucilage as the primary ingredient in marshmallow production. This transformation led to the emergence of marshmallows as a popular sweet treat.

Influence on Culinary Culture The introduction of marshmallow confections in medieval Europe also left a lasting impact on culinary culture. The association of marshmallows with medicinal properties influenced the perception of marshmallows as a "healthier" candy option. This perception, combined with their soft texture and versatility in cooking, contributed to the continued popularity of marshmallows in various cuisines.

Medicinal Revival In recent years, there has been a resurgence in the use of marshmallows for their medicinal properties. Modern herbalists and alternative medicine practitioners recognize the potential therapeutic benefits of the marshmallow plant, particularly its mucilage. Marshmallow preparations are once again being explored for their soothing and healing effects on the respiratory and digestive systems.

Conclusion

The medieval period in Europe played a critical role in the development and utilization of marshmallow as a medicinal remedy. The roots of the marshmallow plant were boiled to extract a mucilage that was then used to treat respiratory and digestive conditions. Despite the challenges of production and limited accessibility, the practice of using marshmallow for its healing properties endured through the centuries. This historical legacy paved the way for the eventual transformation of marshmallow into the beloved confectionery delight we enjoy today.

The Industrial Revolution and Mass Production

The Industrial Revolution, which began in the late 18th century and continued into the 19th century, had a profound impact on the production of marshmallows and headphones. This period saw significant advancements in technology, manufacturing processes, and the overall organization of industries. As a result, mass production became possible, leading to increased efficiency, lower costs, and greater availability of these products.

Technological Advancements

One of the key drivers of the Industrial Revolution was the development and application of new technologies. Several technological advancements played a crucial role in the mass production of marshmallows and headphones.

In the case of marshmallows, the Industrial Revolution brought improvements in the manufacturing process. Prior to this era, marshmallows were made by hand using labor-intensive methods. However, the introduction of steam-powered machinery allowed for faster and more efficient production. Machines were used to automate the mixing, whipping, and shaping of marshmallow ingredients, which greatly increased production capacity.

Similarly, in the case of headphones, the Industrial Revolution led to the invention and refinement of machinery that facilitated mass production. Early headphones were bulky and handcrafted, but with advancements in engineering and mechanization, these devices became more streamlined and affordable. The development of specialized machinery for cutting, molding, and assembling headphone components enabled manufacturers to produce them on a larger scale.

Manufacturing Processes

The Industrial Revolution also brought significant changes to the manufacturing processes involved in producing marshmallows and headphones. Mass production techniques were implemented to increase efficiency and reduce costs.

In the case of marshmallows, the industrialization of their production involved the standardization of ingredients and recipes. This allowed for a consistent quality of marshmallows to be produced in large quantities. The introduction of assembly lines further enhanced efficiency by dividing the production process into sequential tasks performed by specialized workers. This division of labor streamlined the production process, enabling manufacturers to produce marshmallows at a much faster rate.

Similarly, in the case of headphones, the adoption of assembly line production revolutionized the industry. Each worker focused on a specific task, such as soldering wires, attaching ear pads, or assembling drivers. This division of labor not only increased production speed but also allowed for better quality control as each worker became highly skilled in their respective task. Additionally, new materials and manufacturing techniques, such as injection molding for plastic components, were developed during this period, further reducing production costs and making headphones more accessible to a wider audience.

Impact on Society

The mass production of marshmallows and headphones during the Industrial Revolution had a significant impact on society. These products became more affordable and accessible, leading to changes in consumption patterns and cultural practices.

With marshmallows becoming easier to produce and cheaper to buy, they became a popular treat among the general population, no longer reserved for the elite. Marshmallows became synonymous with childhood, and their consumption became intertwined with cultural practices, such as roasting them over campfires or using them in various desserts. The mass production of marshmallows allowed for the creation of new flavors and varieties, catering to different consumer preferences.

Similarly, the mass production of headphones made them more accessible to the general public. Prior to the Industrial Revolution, headphones were primarily used by professionals, such as telephone operators and radio broadcasters. However, with the advent of mass production, headphones became an affordable commodity available to a broader market. This led to a shift in cultural practices, as individuals began using headphones for personal listening and entertainment purposes. The widespread availability of headphones also contributed to the emergence of new genres of music, such as portable audio devices like the Walkman, which revolutionized the way people listened to music.

Overall, the Industrial Revolution and the subsequent mass production of marshmallows and headphones had a profound impact on society. The advancements in technology and manufacturing processes made these products more accessible and affordable to the general public, changing cultural practices and consumption patterns. The legacy of the Industrial Revolution can still be seen today in the widespread availability and variety of marshmallows and headphones in global markets.

Unconventional Example: The Influence on Music Culture

An unconventional but relevant aspect of the Industrial Revolution's impact on mass production and its influence on headphones is the role it played in shaping music culture. With the availability of affordable headphones, individuals gained the ability to listen to music privately and on-the-go. This led to the emergence of music as a more personal and introspective experience.

The newfound accessibility of headphones enabled individuals to explore different genres of music and create personalized playlists. This had a profound impact on how people consumed music and influenced their musical preferences.

As a result, a shift in music culture occurred, with the rise of niche and sub-genres becoming more prominent.

Furthermore, the portability of headphones allowed individuals to take their music with them wherever they went. This led to the growth of the music industry, with artists and record labels capitalizing on the increased demand for portable music. This, in turn, led to the development of new marketing strategies, such as targeted advertising campaigns for specific genres or artists.

In conclusion, the Industrial Revolution and the mass production it facilitated played a significant role in shaping music culture through the accessibility and affordability of headphones. The ability to listen to music privately and on-the-go allowed for a more personal and intimate connection with music. This influence continues to impact music culture, with the availability of headphones shaping the way people consume and interact with music.

Marshmallows in the Modern Era

Marshmallows in Popular Culture

Marshmallows have permeated popular culture in various ways, becoming a beloved symbol that evokes nostalgia, comfort, and whimsy. From movies and TV shows to music and fashion, marshmallows have made their mark and become an integral part of our cultural landscape. In this section, we will explore the diverse presence of marshmallows in popular culture and examine how they have captivated audiences across different mediums.

Marshmallows in Films and Television

Marshmallows have played prominent roles in films and television, often serving as catalysts for comedic or heartwarming moments. In the 1984 film "Ghostbusters," the Stay Puft Marshmallow Man becomes an iconic character, towering over the city and wreaking havoc. This larger-than-life marshmallow figure has become synonymous with the film and continues to be referenced in popular culture. The movie showcases the power of marshmallows to transcend their humble confectionery origins and become a larger symbol of chaos and destruction.

In the television show "Stranger Things," marshmallows take on a different role, becoming a motif that represents innocence and friendship. The main characters often gather around a campfire, toasting marshmallows as a way to bond and find solace in the face of supernatural events. The marshmallow roasting scenes not only provide nostalgic moments but also symbolize the characters' camaraderie and resilience in the face of adversity.

Marshmallows in Music

Marshmallows have also found their way into the realm of music, with numerous songs referencing or using marshmallows as metaphors. In the hit song "Blank Space" by Taylor Swift, she playfully asks her lover if they want to "grab some marshmallows and watch the fire." Here, the mention of marshmallows signifies a cozy and intimate moment, creating a sense of warmth and comfort.

Additionally, in the song "Campfire" by Childish Gambino, marshmallows are lyrically woven into a narrative about love and longing. The lyrics describe the act of roasting marshmallows as a metaphor for the sparks and intensity of a romantic relationship. These musical references demonstrate the ability of marshmallows to evoke emotions and add depth to lyrics, further embedding them into popular culture.

Marshmallows in Fashion and Design

Marshmallows have also made their way into the world of fashion and design, becoming a trendy and whimsical motif. The marshmallow's soft and playful aesthetic has inspired fashion designers to incorporate its shape and colors into clothing, accessories, and even home decor.

Designers have created marshmallow-inspired handbags, sneakers, and even jewelry, featuring fluffy textures and pastel hues. The prominence of marshmallows in fashion showcases their appeal as a symbol of comfort, fun, and indulgence.

The Unconventional Marshmallow Challenge

To further explore the influence of marshmallows in popular culture, let's engage in a creative exercise called the Unconventional Marshmallow Challenge. Gather a group of friends and provide them with a variety of craft materials, such as toothpicks, straws, and tape, along with a bag of marshmallows. In teams, challenge your friends to build the tallest structure using only these materials and marshmallows as connectors.

This activity not only fosters teamwork and creativity but also allows participants to explore the unconventional properties of marshmallows as a structural component. Marshmallows, known for their softness and pliability, present a unique challenge in engineering stable structures. Through this fun and interactive challenge, participants can gain a deeper appreciation for the versatility of marshmallows and their potential beyond their delicious taste.

Conclusion

The presence of marshmallows in popular culture reflects their widespread appeal and ability to evoke emotions and memories. From their appearances in films and television shows to their influence in music, fashion, and even creative challenges, marshmallows have become a beloved symbol of comfort, friendship, and whimsy. Their cultural significance extends beyond their role as a sweet treat, making them an enduring part of our collective imagination.

Marshmallows as a Symbol of Childhood

Marshmallows hold a special place in our hearts as a symbol of childhood. The soft and fluffy nature of marshmallows, combined with their sweet and sugary taste, evokes feelings of nostalgia and innocence. In this section, we will explore why marshmallows have become synonymous with childhood and how they have become an integral part of our cultural and emotional experiences.

The Joy of Marshmallow Consumption

One of the main reasons why marshmallows are associated with childhood is the sheer joy of consumption. From the moment we take a bite of a marshmallow, we are transported to a world of pure delight. The soft texture of marshmallows appeals to our senses, creating a sensory experience that is both comforting and pleasurable. As children, we often enjoyed marshmallows in a variety of forms - roasted over a campfire, melted in hot chocolate, or simply eaten straight from the bag. These experiences create lasting memories and associations with the carefree days of our youth.

Marshmallows in Play and Imagination

Marshmallows also play a significant role in children's play and imagination. Their versatility allows them to be molded into various shapes and structures, encouraging creativity and exploration. Whether building a marshmallow tower or crafting a marshmallow sculpture, children are able to engage in hands-on activities that stimulate their imagination and problem-solving skills. The malleable nature of marshmallows provides a sensory experience that is both captivating and enjoyable for children of all ages.

Marshmallows as a Comfort Food

In addition to their role in play, marshmallows are often seen as a comfort food. Many of us have fond memories of enjoying a warm cup of hot chocolate with a handful of marshmallows on a cold winter's day. The act of savoring a marshmallow brings about a sense of warmth and security, providing solace during times of stress or emotional upheaval. As children, having a marshmallow treat offered comfort and reassurance, creating a sense of safety and well-being.

Marshmallows and Intergenerational Bonding

Marshmallows also serve as a bridge between generations, connecting children with their parents and grandparents. Passing down traditions such as roasting marshmallows over a campfire or sharing a favorite marshmallow recipe creates a sense of continuity and shared experiences. These intergenerational connections strengthen family ties and reinforce the importance of preserving childhood memories.

Cultural Influences on Marshmallows as a Symbol of Childhood

Marshmallows as a symbol of childhood are not limited to a specific culture or region. They have permeated various cultural traditions and have become a universal representation of innocence and joy. For example, in North America, marshmallows are associated with camping and outdoor activities, where families gather around a campfire to share stories and roast marshmallows. In European cultures, marshmallows are often used in festive celebrations, such as the French tradition of making marshmallow-based treats during Easter.

Conclusion

In conclusion, marshmallows hold a special place in our collective memory as a symbol of childhood. Their softness, sweetness, and versatility make them the perfect representation of innocence, joy, and imagination. Whether consumed as a simple treat or used as a building material in creative play, marshmallows have a way of connecting us to our childhood experiences and bringing a sense of comfort and nostalgia. So the next time you bite into a marshmallow, take a moment to appreciate the deeper meaning behind this simple yet profound symbol of childhood.

Marshmallow Consumption and Preferences

Marshmallows have long been enjoyed as a sweet treat by people of all ages. In this section, we will explore the consumption patterns and preferences surrounding marshmallows, including factors that influence the choices individuals make when selecting and consuming marshmallows.

Factors Influencing Marshmallow Consumption

Several factors come into play when it comes to marshmallow consumption. Understanding these factors can shed light on why people are drawn to marshmallows in the first place and how their preferences may differ.

Taste and Texture One of the primary reasons people consume marshmallows is because of their unique taste and texture. The soft, fluffy, and slightly chewy texture combines with a sweet and creamy flavor, making marshmallows a delightful treat for many. The sensation of biting into a marshmallow and feeling it melt in your mouth is a sensory experience that appeals to many individuals.

Availability and Accessibility The availability and accessibility of marshmallows also play a significant role in their consumption patterns. Marshmallows are widely available in grocery stores, convenience stores, and online retailers, making them easily accessible to a wide range of consumers. The affordable price point of marshmallows also contributes to their popularity and widespread consumption.

Cultural and Traditional Practices Marshmallows hold cultural and traditional significance in various societies. For example, marshmallows are often associated with camping trips and outdoor activities where they are toasted over a campfire and used in the creation of s'mores. These cultural practices contribute to the consumption of marshmallows in specific contexts.

Marshmallow Preferences

While marshmallows may have a universal appeal, individual preferences can vary. Here are some factors that influence marshmallow preferences among consumers.

Flavor Varieties Marshmallows are available in a variety of flavors beyond the classic vanilla. Some popular flavor variations include strawberry, chocolate, toasted coconut, and peppermint. Different individuals have different flavor preferences, and the availability of various options allows consumers to choose their preferred flavors.

Usage and Application Marshmallows can be enjoyed in various ways, and this can impact individual preferences. Some individuals prefer to consume marshmallows on their own, while others enjoy incorporating them into recipes or using them as ingredients in desserts. The versatility of marshmallows allows individuals to tailor their consumption to their specific tastes and preferences.

Health Considerations Health considerations can also influence marshmallow preferences. Some individuals are mindful of their sugar intake and may prefer sugar-free or low-sugar marshmallow options. Additionally, individuals with dietary restrictions or preferences, such as vegans or individuals with gluten intolerance, may seek out marshmallows that cater to their specific needs.

Case Study: The Rise of Gourmet Marshmallows

In recent years, there has been a surge in the popularity and consumption of gourmet marshmallows. These upscale marshmallows offer unique flavors, premium ingredients, and artisanal craftsmanship. This trend highlights the evolving preferences of consumers who are seeking a more sophisticated and innovative marshmallow experience.

Gourmet marshmallows often come in flavors such as lavender, salted caramel, champagne, and even savory options like rosemary and bacon. They are frequently used in gourmet desserts, hot beverages, and as standalone treats.

The rise of gourmet marshmallows can be attributed to several factors. Firstly, it taps into the growing consumer demand for unique and artisanal food experiences. These marshmallows offer a departure from the traditional flavors and textures, providing a gourmet twist on a beloved classic.

Secondly, gourmet marshmallows cater to individuals who prioritize quality ingredients and artisanal production methods. Many gourmet marshmallow brands use natural flavors, organic ingredients, and small-batch manufacturing processes. This focus on quality and craftsmanship appeals to consumers seeking an elevated marshmallow experience.

Lastly, the market for gourmet marshmallows has been nurtured by social media platforms and food blogs. Gourmet marshmallows are Instagrammable, and people love sharing visually appealing food experiences online. This exposure has contributed to the popularity and increased consumption of gourmet marshmallows.

Conclusion

Marshmallow consumption and preferences are influenced by several factors, including taste and texture, availability, cultural practices, and individual preferences. Understanding these factors can help manufacturers and marketers cater to the diverse consumer base and create innovative marshmallow products that align with changing consumer preferences. The rise of gourmet marshmallows serves as a testament to the ever-evolving nature of marshmallow consumption, providing consumers with new and exciting options to explore. As we continue to explore the global perspectives on marshmallows, it is essential to keep in mind the dynamic nature of consumer preferences and the opportunities for further innovation in marshmallow production and consumption.

Health and Nutritional Considerations of Marshmallows

When it comes to marshmallows, most people don't immediately think about health and nutrition. After all, these fluffy confections are usually associated with sweet treats and indulgence. However, it's important to consider the impact that marshmallows can have on our bodies and overall well-being. In this section, we will explore the health and nutritional aspects of marshmallows, taking into account their ingredients, caloric content, and potential effects on our health.

Ingredients in Marshmallows

To understand the health implications of marshmallows, we first need to take a closer look at their ingredients. Traditional marshmallows are made primarily from sugar, corn syrup, and gelatin, along with other flavorings and stabilizers. Let's break down these ingredients and examine their nutritional profiles:

- **Sugar:** Sugar is the main component of marshmallows and contributes to their sweet taste. However, excessive consumption of sugar has been linked to various health issues, including obesity, type 2 diabetes, and dental cavities. It's important to consume sugar in moderation to maintain a healthy diet.

- **Corn syrup:** Corn syrup is used in marshmallows to provide sweetness and aid in their texture. It is derived from cornstarch and is a source of simple carbohydrates. High consumption of corn syrup has been associated with an increased risk of obesity and metabolic disorders. It's advisable to limit our intake of corn syrup and opt for healthier alternatives whenever possible.

- **Gelatin:** Gelatin is a protein obtained from animal collagen and is responsible for the unique texture and structure of marshmallows. While gelatin itself is not harmful, it may not be suitable for individuals following certain dietary restrictions, such as vegetarians or those who avoid animal byproducts.

- **Flavorings and stabilizers:** Additional ingredients, such as vanilla extract or food colorings, may be used to enhance the flavor and appearance of marshmallows. While these additives are generally safe, some artificial food colorings have been linked to hyperactivity in children and may cause allergies or sensitivities in some individuals. It's important to be aware of these potential side effects and choose marshmallows with natural or plant-based ingredients if desired.

Caloric Content and Portion Size

Marshmallows are relatively low in calories compared to many other sweet treats. On average, a single regular-sized marshmallow contains around 25 calories. However, it's crucial to consider portion sizes when consuming marshmallows, as excessive intake can contribute to weight gain and other health issues.

A typical serving size of marshmallows is around four marshmallows, which equates to approximately 100 calories. It's important to be mindful of portion sizes and avoid overindulging, especially if you are watching your calorie intake.

Health Impact

While marshmallows may not be a health food, they can still be enjoyed in moderation as part of a balanced diet. Here are some key factors to consider regarding the health impact of marshmallows:

- **Sugar intake:** Marshmallows are a concentrated source of sugar, and excessive consumption can lead to negative health effects. It's important to be mindful of your overall sugar intake and limit your consumption of foods high in added sugars.

- **Nutrient deficiencies:** Marshmallows are low in essential nutrients such as vitamins, minerals, and fiber. Relying on marshmallows as a significant part of your diet may result in nutrient deficiencies. It's important to prioritize nutrient-rich foods to meet your dietary needs.

- **Weight management:** Due to their high sugar content and low nutritional value, consuming marshmallows in excess can contribute to weight gain and hinder weight management efforts. It's advisable to consume marshmallows sparingly and opt for healthier snacks that provide more nutrients and fiber.

- **Dental health:** Marshmallows, like other sugary foods, can contribute to tooth decay and dental cavities. The sugary residue left on teeth after consuming marshmallows can create an environment for bacteria to thrive. It's crucial to practice proper oral hygiene, including regular brushing and flossing, to maintain good dental health.

Alternative Options and Moderation

If you enjoy the taste and texture of marshmallows but want to make healthier choices, there are alternatives available. Consider the following options:

- **Homemade marshmallows:** Making marshmallows at home allows you to control the ingredients and customize the recipe to suit your preferences. You can experiment with healthier alternatives to sugar and corn syrup, such as honey or maple syrup, and even try using plant-based gelatin substitutes like agar agar.

- **Natural or organic marshmallows:** Look for marshmallows made with natural or organic ingredients. These options often have fewer artificial additives and may use plant-based alternatives to gelatin. However, it's still important to check the nutritional label and consume them in moderation.

- **Marshmallow alternatives:** If you're looking for a healthier alternative to traditional marshmallows, consider options like marshmallow root tea or marshmallow-flavored desserts made with natural ingredients. These alternatives can satisfy your craving for marshmallow flavor without the added sugar and artificial ingredients.

Conclusion

While marshmallows may not be the healthiest food choice, they can still be enjoyed in moderation as part of a balanced diet. Being mindful of portion sizes, considering healthier alternatives, and prioritizing nutrient-rich foods are key to maintaining a healthy lifestyle. By understanding the health and nutritional considerations of marshmallows, you can make informed choices and enjoy these sweet treats responsibly.

The Future of Marshmallows

Trends in Marshmallow Flavors and Varieties

Marshmallows have come a long way from their humble beginnings as a confection made from the root of the marshmallow plant. Today, marshmallows are available in a wide array of flavors and varieties that cater to the ever-changing tastes and preferences of consumers. In this section, we will explore some of the current trends in marshmallow flavors and varieties, as well as the factors driving these trends.

The Rise of Artisanal Marshmallows

One notable trend in marshmallow flavors and varieties is the growing popularity of artisanal marshmallows. Artisanal marshmallows are made in small batches using high-quality ingredients and often feature unique and innovative flavors. These marshmallows are not only known for their superior taste but also for their visually appealing appearance.

Artisanal marshmallow flavors often go beyond the traditional vanilla and include options such as salted caramel, dark chocolate, lavender, matcha green tea, and even bacon. These unconventional flavors have gained a strong following among food enthusiasts who seek out novel and exciting taste experiences.

The trend towards artisanal marshmallows can be attributed to the increasing demand for gourmet and handmade food products. Consumers are becoming more adventurous in their culinary choices and are willing to pay a premium for high-quality, unique marshmallow flavors.

Health-Conscious Marshmallows

As the health and wellness movement continues to gain momentum, there is a growing demand for healthier alternatives to traditional marshmallows. Health-conscious consumers are seeking marshmallows that are free from artificial ingredients, preservatives, and excessive sugar.

To cater to this market, manufacturers have started producing marshmallows sweetened with natural sweeteners like honey, maple syrup, and coconut sugar. These alternatives provide a more wholesome option for those who want to indulge in marshmallows without compromising their dietary preferences.

Furthermore, there has been a rise in marshmallow varieties that are gluten-free, dairy-free, and vegan-friendly. These options allow individuals with specific dietary restrictions or preferences to enjoy marshmallows without any concerns.

Innovative Texture and Shape

In addition to new flavors, marshmallow varieties are also adopting innovative textures and shapes. Manufacturers are experimenting with different textures to create marshmallows that are softer, chewier, or even crispy. These variations offer a unique sensory experience and add a new dimension to the enjoyment of marshmallows.

Furthermore, marshmallows are now available in various shapes and sizes beyond the traditional cylindrical form. Some examples include mini marshmallows, giant marshmallows, heart-shaped marshmallows, and even character-shaped marshmallows. These creative shapes not only make marshmallows visually appealing but also add an element of fun and playfulness.

Cross-cultural Influences

As global cultures continue to blend and influence each other, the flavors and varieties of marshmallows are also being shaped by cross-cultural exchanges. Manufacturers are incorporating flavors and ingredients from different cuisines and traditions to create unique marshmallow experiences.

For example, Asian-inspired flavors like matcha, lychee, and yuzu have found their way into marshmallow varieties. Similarly, European flavors such as tiramisu, strawberry cheesecake, and champagne are being incorporated into marshmallow creations. This cross-pollination of flavors adds diversity and excitement to the world of marshmallows.

The Future of Marshmallow Flavors and Varieties

Looking ahead, the trends in marshmallow flavors and varieties are likely to continue evolving. With advancements in food science and technology, manufacturers will have even more options to experiment with new flavors, textures, and shapes.

Furthermore, consumer preferences for healthier and more sustainable food options will continue to shape the development of marshmallow varieties. We can expect to see an increase in marshmallows that are low in sugar, made with natural and organic ingredients, and cater to specific dietary needs.

In conclusion, the world of marshmallows is undergoing a transformation with an abundance of exciting and innovative flavors and varieties. From artisanal gourmet marshmallows to health-conscious alternatives, there is a marshmallow for every taste and preference. With the continued exploration and experimentation in this field, we can look forward to even more delightful surprises in the future. So go ahead, explore the vast world of marshmallows and indulge in the sweet pleasures they offer.

Innovations in Marshmallow Manufacturing

In recent years, the marshmallow industry has witnessed several exciting innovations in manufacturing techniques. These advancements have enhanced the quality, flavor, and variety of marshmallows, while also improving efficiency and sustainability in the production process. This section will explore some of the key innovations that have shaped the modern marshmallow manufacturing industry.

Gelatin Substitutes

Traditionally, marshmallows were made using gelatin, which is derived from animal collagen. However, with the increasing demand for vegetarian and vegan options, manufacturers have sought alternative ingredients to replace gelatin. One of the most notable innovations in marshmallow manufacturing is the development of gelatin substitutes that provide similar texture and binding properties.

Agar-agar, a substance derived from seaweed, has emerged as a popular gelatin substitute in marshmallow production. It offers a similar gelling effect and can be used in equal quantities as gelatin. Other alternatives include carrageenan, a natural extract from red seaweed, and pectin, a plant-based gelling agent derived from fruits.

These gelatin substitutes not only cater to dietary preferences but also contribute to environmental sustainability by reducing the reliance on animal-derived ingredients.

Flavor Infusions

Innovation in marshmallow manufacturing has also led to a wide range of exciting flavor options. Manufacturers have developed innovative techniques to infuse marshmallows with various flavors, ranging from classic favorites to more unconventional and exotic choices.

One method involves incorporating natural fruit extracts into the marshmallow mixture during the cooking process. This allows for the creation of fruit-flavored marshmallows without the need for artificial additives. For example, strawberry, raspberry, and lemon-flavored marshmallows have become popular alternatives to the traditional vanilla flavor.

Another innovative approach is the use of infusion technology. This technique involves infusing the marshmallow mixture with flavors through a process of soaking or steeping. For instance, marshmallows infused with flavors like lavender, chai tea, or even bacon have gained popularity among adventurous consumers.

These flavor innovations provide consumers with a wider selection of tastes and cater to their changing preferences while maintaining the classic marshmallow texture and melt-in-your-mouth experience.

Texture Modification

Texture plays a crucial role in the overall sensory experience of marshmallows. Innovations in texture modification techniques have enhanced the mouthfeel of marshmallows, offering consumers unique and satisfying experiences.

One technique used to modify the texture of marshmallows is the incorporation of microbubbles. By introducing tiny air bubbles into the marshmallow mixture, manufacturers can achieve a lighter and fluffier texture. This has proven especially popular for marshmallows used as dessert toppings or in confectionery.

Another texture modification innovation is the use of additives, such as maltodextrin or tapioca starch. These additives absorb moisture, resulting in a drier and less sticky marshmallow texture. This modification is especially useful for marshmallows intended for baking or as ingredients in other food products.

These texture modifications not only diversify the marshmallow eating experience but also open up new applications for marshmallows in culinary creations.

Production Automation

Automation has played a significant role in improving efficiency and consistency in marshmallow manufacturing. Innovations in production processes have reduced human labor and increased productivity while maintaining the quality standards of marshmallow production.

One key innovation is the automation of mixing and cooking processes. Automated systems ensure precise measurements of ingredients, resulting in consistent marshmallow texture and taste. Additionally, automated temperature control during the cooking process ensures that the marshmallow mixture reaches the desired consistency and avoids scorching or burning.

Another area of automation is the cutting and shaping of marshmallows. Automated cutting machines can cut marshmallow slabs or logs into uniform shapes and sizes, reducing waste and increasing production speed.

The integration of robotics and artificial intelligence into marshmallow manufacturing has further enhanced efficiency and precision. Robots can perform repetitive tasks with remarkable accuracy and speed, freeing up human workers to focus on more complex aspects of production and quality control.

Sustainability Initiatives

As consumer awareness about sustainability increases, the marshmallow industry has responded with innovative initiatives to reduce its environmental impact.

One such innovation is the use of sustainable packaging materials. Manufacturers have started incorporating biodegradable and compostable materials for marshmallow packaging, such as plant-based films or recycled paper. This reduces the environmental footprint associated with packaging waste.

Additionally, manufacturers have implemented energy-efficient production processes. This includes the use of advanced heating and cooling systems that minimize energy consumption, as well as the installation of renewable energy sources, such as solar panels, to power the manufacturing facilities.

Efforts have also been made to optimize water usage in marshmallow production. Innovative water recycling and purification systems help reduce water consumption and minimize the release of pollutants into the environment.

By embracing sustainability initiatives, the marshmallow industry can not only reduce its environmental impact but also meet the growing demand for eco-friendly products from conscious consumers.

Conclusion

Innovations in marshmallow manufacturing have revolutionized the industry in recent years. Gelatin substitutes, flavor infusions, texture modifications, production automation, and sustainability initiatives have all contributed to the growth, variety, and sustainability of the marshmallow market.

These innovations have not only catered to changing consumer preferences but have also driven economic growth, improved production efficiency, and reduced the environmental impact of marshmallow manufacturing.

As the marshmallow industry continues to evolve, it is essential for manufacturers to prioritize innovation, sustainability, and consumer-centric approaches to ensure the continued success and popularity of marshmallows in the global market.

By keeping up with technological advancements and consumer demands, the marshmallow industry can continue to thrive, delighting taste buds and sparking joy worldwide.

Potential Impacts of Climate Change on Marshmallow Production

Climate change is a pressing issue that affects various aspects of our lives, including the production of marshmallows. Marshmallow production relies on a stable and suitable climate throughout the growth and harvesting process of the key ingredient, gelatin. In this section, we will explore the potential impacts of climate change on the production of marshmallows and discuss the measures that can be taken to mitigate its effects.

Understanding Climate Change

Before delving into the specific impacts, let's first understand the concept of climate change. Climate change refers to long-term shifts in weather patterns and average temperatures of a given region. It is primarily caused by the accumulation of greenhouse gases, such as carbon dioxide and methane, in the Earth's atmosphere. These gases trap heat and contribute to the warming of the planet—a phenomenon known as the greenhouse effect.

Human activities, including the burning of fossil fuels, deforestation, and industrial processes, have significantly increased the concentration of greenhouse gases in the atmosphere, exacerbating the effects of climate change. The consequences of climate change are far-reaching, affecting ecosystems, weather patterns, agriculture, and many other sectors.

Impacts on Marshmallow Production

1. **Changes in growing seasons:** Marshmallow plants require specific temperature and moisture conditions for optimal growth. Climate change can disrupt these ideal conditions, leading to alterations in the duration and timing of growing seasons. Higher temperatures and shifting precipitation patterns can affect the germination, flowering, and maturation of marshmallow plants. This, in turn, can impact the yield and quality of the gelatin extracted from the plants.

2. **Water availability:** Marshmallow plants require an adequate water supply for growth. However, climate change can cause changes in precipitation patterns, leading to droughts or excessive rainfall in certain regions. Droughts can reduce water availability for irrigation, affecting the productivity of marshmallow farms. Conversely, heavy rainfall can result in waterlogging and plant diseases, further compromising the crop.

3. **Increased pest and disease pressure:** Climate change can alter the distribution and population dynamics of pests and diseases that affect marshmallow plants. Warmer temperatures can create favorable conditions for the

proliferation of pests, such as aphids and mites, which can damage marshmallow crops. Changes in precipitation patterns can also contribute to the spread of fungal or bacterial diseases. As a result, farmers may face increased challenges in pest and disease management, potentially reducing marshmallow yields.

4. **Extreme weather events:** Climate change is associated with an increased frequency and intensity of extreme weather events, such as hurricanes, heatwaves, and heavy storms. These events can cause physical damage to marshmallow farms, including crop destruction, soil erosion, and infrastructure loss. Such disruptions can significantly impact the production and supply of marshmallows, leading to market instability and higher prices.

Mitigation and Adaptation Strategies

To address the potential impacts of climate change on marshmallow production, several strategies can be employed:

1. **Crop diversification:** Farmers can explore the cultivation of alternative crops or varieties that are more resilient to changing climatic conditions. Diversifying the crops can help reduce the dependence on marshmallows and mitigate the risks associated with climate change.

2. **Improving water management:** Implementing efficient irrigation systems and water conservation practices can help ensure the availability of water for marshmallow cultivation, even in the face of changing precipitation patterns. Techniques such as drip irrigation and water recycling can minimize water wastage and enhance overall water-use efficiency.

3. **Integrated pest management:** Adopting integrated pest management practices can help control pests and diseases without relying heavily on chemical pesticides. This approach involves monitoring pest populations, implementing cultural and biological control methods, and using chemical control only when necessary. By reducing chemical inputs, farmers can minimize environmental pollution and promote a more sustainable production system.

4. **Investing in research and development:** Continued research and development efforts are crucial to developing climate-resilient varieties of marshmallow plants. Breeding programs can focus on improving heat and drought tolerance, disease resistance, and overall productivity. Additionally, research can also explore innovative cultivation techniques that minimize the environmental footprint of marshmallow production.

Case Study: Marshmallow Farms in California

The state of California in the United States is known for its marshmallow production. However, the region has been experiencing the effects of climate change, including prolonged droughts and heatwaves. These changes have posed significant challenges for marshmallow farmers, affecting both their yields and profitability.

To mitigate these impacts, California farmers have implemented various adaptation strategies:

1. **Transitioning to drip irrigation:** Traditional flood irrigation methods have been replaced with more efficient drip irrigation systems. By delivering water directly to the root zone of the plants, drip irrigation minimizes water wastage and increases water-use efficiency.

2. **Crop rotation and cover cropping:** Farmers have implemented crop rotation practices to maintain soil health and reduce pest and disease pressure. Additionally, cover crops, such as legumes, are grown to improve soil fertility and reduce erosion.

3. **Investing in climate data and forecasting:** Marshmallow farmers in California are utilizing climate data and weather forecasting to make informed decisions about irrigation scheduling, pest control, and other farm management practices. This helps optimize resource utilization and minimize the risks associated with climate variability.

These adaptation strategies highlight the importance of proactive measures in mitigating the potential impacts of climate change on marshmallow production. By embracing sustainable practices and harnessing scientific advancements, farmers can ensure the long-term viability of the industry.

Conclusion

Climate change poses significant challenges to the production of marshmallows, a globally popular confectionery. Alterations in temperature, water availability, pest dynamics, and extreme weather events can disrupt the growth and harvest of marshmallow crops, impacting their quality and quantity. However, through strategic mitigation and adaptation efforts, such as crop diversification, improved water management, integrated pest management, and research and development, the negative impacts of climate change on marshmallow production can be minimized. By prioritizing sustainable practices and embracing technological advancements, marshmallow farmers can navigate the evolving climate landscape and continue to meet the global demand for this beloved treat.

In the next chapter, we will shift our focus to the evolution of headphones and explore the technological advancements and cultural significance of these audio devices.

Chapter 2: The Evolution of Headphones

Early Sound Reproduction Devices

The Development of Ear Trumpets and Stethoscopes

In this section, we will explore the fascinating history of two important sound reproduction devices: ear trumpets and stethoscopes. These devices played a crucial role in the development of headphones as we know them today. We will delve into their origins, their impact on medical practices, and their influence on the subsequent invention of headphones.

Origins and Early Uses

The concept of the ear trumpet can be traced back to ancient times, with evidence of similar devices being used by various civilizations. However, it was not until the 17th century that the first documented designs for ear trumpets emerged.

During this time, people with hearing loss relied on ear trumpets to amplify sound. The early designs were cone-shaped devices made from materials such as wood, metal, or even animal horns. These devices worked by capturing sound waves and funneling them into the user's ear, increasing the volume and clarity of the sound.

In the 19th century, the stethoscope was invented by French physician René Laennec. Laennec found it difficult to perform traditional auscultation (listening to internal sounds of the body) due to the social norms of the time, particularly the need to maintain distance from female patients. This led him to develop a hollow wooden cylinder, which he called a stethoscope (from the Greek words stethos, meaning "chest," and skopein, meaning "to examine"). With the stethoscope, Laennec could listen to sounds emitted by the chest and abdomen, aiding in the diagnosis of various medical conditions.

Advances in Design and Functionality

As the demand for ear trumpets and stethoscopes grew, so did the need for improved designs and functionalities. Technological advancements and evolving medical practices played a significant role in the development of these devices.

In the case of ear trumpets, the cone-shaped design was refined to enhance sound amplification. Manufacturers experimented with different materials, such as brass and silver, to improve the transmission of sound waves. Additional features, such as adjustable tubes and earpieces, were also incorporated to enhance comfort and usability.

Similarly, stethoscopes underwent several design iterations to improve their effectiveness. Initially, the stethoscope consisted of a single tube directly applied to the patient's body. However, this design posed limitations, particularly when examining hard-to-reach areas or during surgery. To overcome these challenges, double-tube and binaural stethoscopes were developed, allowing physicians to listen to internal sounds with greater precision and flexibility.

Impact on Headphone Invention

The development of ear trumpets and stethoscopes set the stage for the invention of headphones. These early sound reproduction devices provided valuable insights into the mechanics of sound transmission and perception. Moreover, they sparked curiosity and innovation in the field of sound reproduction.

The concept of using separate sound receivers placed over or in the ears, as seen in ear trumpets and stethoscopes, laid the foundation for the design of headphones. Inventors and engineers recognized the potential of these devices not only in medical settings but also in areas such as telephony, broadcasting, and music production.

The evolution from ear trumpets and stethoscopes to modern headphones utilized advancements in material science, acoustic engineering, and electrical technology. Today, headphones have become an integral part of our daily lives, providing us with immersive audio experiences, private listening, and communication.

Conclusion

The development of ear trumpets and stethoscopes paved the way for the invention of headphones. These early sound reproduction devices not only amplified sound but also sparked curiosity and innovation in the field of sound transmission and perception.

Through improvements in design and functionality, ear trumpets and stethoscopes enhanced our understanding of how we perceive and reproduce sound. Their impact extended beyond medical practices, influencing the evolution of headphones and their use in various industries.

As we explore the history of marshmallows and headphones in this book, it is important to recognize the contributions and influences of these early sound reproduction devices. The development of ear trumpets and stethoscopes represents a crucial chapter in the global perspectives on marshmallows and headphones.

The Invention of the Telephone and Earpieces

The invention of the telephone and earpieces revolutionized communication and paved the way for the development of modern headphones. In this section, we will explore the origins of the telephone and the early development of earpieces, highlighting their impact on the evolution of headphones.

Origins of the Telephone

The invention of the telephone is credited to Alexander Graham Bell, a Scottish-born scientist and inventor. Bell's fascination with sound and speech led him to experiment with transmitting sound signals over a wire. On March 10, 1876, Bell successfully transmitted the first intelligible phrase over a telephone to his assistant, Thomas Watson, with the famous words, "Mr. Watson, come here. I want to see you."

Bell's telephone was based on the principle of converting sound waves into electrical signals and vice versa. He achieved this by using a diaphragm, an electromagnetic coil, and a permanent magnet. When sound waves hit the diaphragm, it vibrated, causing the electromagnetic coil to move in response, thus generating an electrical signal. This signal was then transmitted over a wire to a receiving end, where it was converted back into sound using a similar setup.

Early Earpieces

In the early days of telephony, conversations were conducted using a two-piece setup, where both the sender and receiver had to hold a separate earpiece and microphone. These early earpieces, also known as receivers, were simple devices designed to convert electrical signals back into audible sound. They consisted of a small speaker housed in a wooden or metal casing, which was held up to the user's ear.

One significant advancement in earpiece design came with the invention of the carbon microphone by Thomas Edison in 1877. The carbon microphone improved the clarity and quality of the transmitted voice signal, making telephone conversations more intelligible. This advancement led to the development of more efficient and compact earpieces, as the improved sound quality allowed for smaller speakers to be used.

Impact on the Evolution of Headphones

The invention of the telephone and the early development of earpieces laid the foundation for the evolution of headphones. As telephony technology advanced, the need for hands-free communication grew, leading to the development of headsets that incorporated both an earpiece and a microphone.

Early telephone headsets were bulky and consisted of a headband that held the earpiece and microphone in place. These headsets allowed users to have conversations without needing to hold a separate earpiece and microphone, providing a more convenient communication experience.

Over time, advancements in miniaturization and material science led to the development of smaller, more lightweight headsets. The introduction of headsets with adjustable headbands and cushioned ear cups improved comfort and usability, making them more practical for extended use.

The evolution of headphones continued with the introduction of stereo sound in the 20th century. With stereo sound, headphones became not only a tool for communication but also a means for enjoying music and other audio content. The development of various sound reproduction technologies, such as dynamic drivers and balanced armature drivers, further improved the audio quality delivered by headphones.

Today, headphones have become an integral part of our lives, with a wide range of styles and options to suit different needs and preferences. From wired headphones to wireless Bluetooth models, noise-canceling technology to smart features, headphones have come a long way since their early telephony roots.

Examples and Applications

The impact of the invention of the telephone and earpieces can be seen in various real-world examples and applications. For instance, in call centers and customer service industries, employees rely on headsets to communicate with customers efficiently and comfortably. These headsets allow for hands-free communication, improving productivity and customer satisfaction.

In the entertainment industry, headphones play a crucial role in music production and recording. Musicians and audio engineers use high-quality headphones to monitor and fine-tune the sound during the recording and mixing process. The accuracy and fidelity of headphones greatly affect the final audio output, making them essential tools in the industry.

Additionally, the development of headsets with built-in microphones has enabled advancements in voice recognition technology and voice-controlled devices. Applications such as virtual assistants and voice commands in smartphones rely on accurate audio input and output, all made possible by the integration of earpieces and microphones in headsets.

Conclusion

The invention of the telephone and the subsequent development of earpieces played a vital role in the evolution of headphones. These technological advancements paved the way for the development of hands-free communication devices, making communication more convenient and accessible. Furthermore, the integration of earpieces and microphones in headsets laid the foundation for the expansion of headphones beyond telephony applications, leading to their widespread use in music, entertainment, and various other industries. The impact of the telephone and earpieces can still be seen in the design and functionality of modern headphones, making them an essential part of our daily lives.

The Emergence of the First Headphones

The emergence of the first headphones marked a significant milestone in the evolution of sound reproduction devices. It revolutionized the way we listen to music, communicate, and experience the world of audio. In this section, we will explore the origins and development of headphones, tracing their evolution from early sound reproduction devices to their modern iterations.

Early Sound Reproduction Devices

To understand the emergence of headphones, we must first delve into the early sound reproduction devices that paved the way for their invention. These devices laid the foundation for the technological advancements that eventually led to the creation of headphones.

The Development of Ear Trumpets and Stethoscopes

Before headphones, there were ear trumpets and stethoscopes. These early devices were designed to amplify sound for individuals with hearing impairments or for medical purposes. Ear trumpets, which resembled large metal funnels, were used to direct sound waves into the ear canal, increasing the volume of sound perceived by the listener. Stethoscopes, on the other hand, were primarily used by medical professionals to listen to internal body sounds and diagnose ailments.

The invention of ear trumpets and stethoscopes laid the groundwork for the concept of directing sound directly into the listener's ears. This concept would later be refined and developed further in the creation of headphones.

The Invention of the Telephone and Earpieces

The invention of the telephone in the late 19th century brought about a new era of communication. With the telephone, sound transmission over long distances became possible. Alexander Graham Bell, as one of the pioneers of the telephone, recognized the need for a device that could deliver sound directly to the listener's ears without the need for a loudspeaker.

To address this need, early telephones were equipped with earpieces connected by a flexible tube to the main body of the telephone. The user would hold the earpiece to their ear to listen to the conversation. Although these early earpieces were not yet true headphones as we know them today, they marked the beginning of a shift towards personal listening devices that would eventually lead to the creation of headphones.

The Emergence of the First Headphones

The emergence of the first headphones came with the recognition of the limitations of earpieces attached to telephones. Engineers and inventors started to explore the possibility of developing a more efficient and comfortable way to deliver sound directly to the listener's ears.

One notable milestone in the development of headphones was the invention of the electrophone in 1910 by Nathaniel Baldwin. The electrophone was one of the earliest examples of a headphone-like device. It consisted of two earpieces connected by a headband, which allowed for hands-free listening. The earpieces were equipped with small electromagnets that produced sound when connected to a power source.

The electrophone gained popularity initially among telephone operators and later among radio operators and early aviators who needed a reliable and convenient way to listen to transmissions. However, due to the manufacturing complexity and high cost, the electrophone remained a niche device and did not achieve widespread adoption.

Contemporary Headphone Technologies

Fast forward to the present, and we find ourselves surrounded by a vast array of headphone technologies. Over the years, headphones have undergone numerous advancements and innovations to meet the ever-increasing demands of consumers for better sound quality, comfort, and convenience.

Wireless and Bluetooth Headphones

Wireless headphones represent a significant leap forward in headphone technology. They eliminate the need for physical connections to the audio source, providing users with freedom of movement and flexibility. Bluetooth technology, in particular, has played a pivotal role in the widespread adoption of wireless headphones. It allows for seamless connectivity between audio devices, opening up a world of possibilities for wireless audio transmission.

Noise-Canceling and Noise-Isolating Headphones

Noise-canceling and noise-isolating headphones have become indispensable tools for those seeking to escape the noise of the outside world and immerse themselves in their music or audio content. Noise-canceling headphones use advanced technology to actively reduce or eliminate ambient noise, while noise-isolating headphones rely on passive sound-blocking mechanisms such as ear cups or in-ear seals to minimize external noise.

Smart Headphones and Wearable Technology

In recent years, the intersection of headphones and wearable technology has given rise to smart headphones. These headphones integrate various features such as biometric sensors, voice assistants, gesture controls, and even augmented reality functionalities. Smart headphones redefine the audio experience by combining traditional audio playback with innovative technologies, opening up new possibilities for audio enthusiasts and tech-savvy individuals.

In conclusion, the emergence of the first headphones can be traced back to the early sound reproduction devices such as ear trumpets, stethoscopes, and telephone earpieces. These early devices laid the foundation for the development of headphones, which have evolved significantly over time. From the invention of the electrophone to the advent of wireless, noise-canceling, and smart headphones, the world of headphones continues to evolve, shaping our audio experiences and enhancing our connectivity in an increasingly digital age.

Headphones in the 20th Century

The Rise of Radio and Broadcasting

The development of radio broadcasting had a profound impact on society, transforming the way people communicate and consume information. In this section, we will explore the historical context, technological advancements, and the cultural significance of the rise of radio and broadcasting.

Historical Context

The origins of radio can be traced back to the late 19th century when inventors such as Nikola Tesla and Guglielmo Marconi experimented with wireless communication. However, it was not until the early 20th century that radio broadcasting became a reality.

One crucial event in the history of radio was the sinking of the RMS Titanic in 1912. The distress signals sent by the ship's wireless operator demonstrated the potential of radio for long-distance communication and emergency broadcasting. Following this tragedy, governments and private companies recognized the need for regulations and infrastructure to harness the power of radio.

Technological Advancements

The development of radio broadcasting required several key technological advancements. The invention of the vacuum tube by Lee De Forest in 1906 was a significant breakthrough. Vacuum tubes allowed for the amplification and modulation of radio signals, leading to the creation of more powerful and reliable broadcasting systems.

Another critical technological development was the improvement of antenna design. Efficient antennas enabled the transmission and reception of radio signals over longer distances, expanding the reach of radio broadcasting.

Furthermore, advancements in audio processing technology, such as the introduction of frequency modulation (FM) by Edwin Armstrong in the 1930s, improved sound quality and reduced interference. FM radio soon surpassed amplitude modulation (AM) radio in popularity due to its superior audio fidelity.

Cultural Significance

The rise of radio and broadcasting had a profound cultural impact worldwide. It revolutionized the way people accessed information and entertainment, bringing news, music, and storytelling into people's homes. Radio became a vital medium for disseminating information during times of crisis and played a significant role in shaping public opinion.

Radio broadcasting also played a critical role in promoting and preserving cultural heritage. It provided a platform for artists, musicians, and writers to showcase their talents and share their works with a broader audience. Moreover, radio allowed people to connect with different cultures and communities through the airwaves, fostering cross-cultural understanding and appreciation.

Radio as a Propaganda Tool

It is essential to acknowledge that radio broadcasting was also utilized as a tool for propaganda during times of war and political unrest. Governments and ideological groups recognized the power of radio to shape public opinion and disseminate their messages.

Examples of this include the Nazi regime's use of radio to spread its propaganda during World War II and the Cold War era, where radio played a crucial role in the ideological battle between the United States and the Soviet Union.

Challenges and Regulation

The rapid growth of radio broadcasting presented several challenges. One of the main challenges was finding available radio spectrum frequencies to accommodate the increasing number of broadcasting stations. This led to the establishment of regulatory bodies like the Federal Communications Commission (FCC) in the United States to allocate and manage the use of radio frequencies.

Another significant challenge was the need for standardized broadcasting formats and protocols. This led to the development of broadcasting systems such as AM and FM, which became the global standards for radio transmission.

Conclusion

The rise of radio and broadcasting revolutionized communication and had a profound impact on society. The technological advancements, cultural significance, and challenges associated with radio broadcasting shaped the media landscape as we know it today.

The development of radio not only provided a means of entertainment but also facilitated the dissemination of information, preservation of cultural heritage, and even served as a tool for propaganda. It was a transformative era that paved the way for further advancements in communication technology.

The Introduction of Stereo Sound and Hi-Fi Systems

The introduction of stereo sound and hi-fi systems revolutionized the way people experienced music in the 20th century. This section explores the development of stereo sound technology and its impact on the audio industry, music production, and the listening habits of consumers.

The Birth of Stereo Sound

Before the invention of stereo sound, audio reproduction was limited to mono or monaural sound, where a single audio channel delivered the entire audio signal to the listener. The stereo sound technology introduced a new dimension to audio playback by creating a sense of depth and spatiality.

In the early 20th century, inventors and engineers began experimenting with ways to reproduce sound using multiple channels. One early attempt was the binaural recording technique, which employed two microphones to capture sound separately and create a stereo effect. However, this method required listeners to wear headphones to experience the full stereo effect.

The breakthrough came in the 1930s with the development of the stereo disc system, commonly known as vinyl records. These records featured two separate channels embedded in the grooves, allowing for the playback of stereo sound on compatible record players. The stereo disc system quickly gained popularity, paving the way for the widespread adoption of stereo sound.

Hi-Fi Systems and Fidelity

The term "hi-fi" stands for high fidelity, referring to audio systems that accurately reproduce sound as close to the original recording as possible. Hi-fi systems became popular in the mid-20th century, offering improved sound quality and a more immersive listening experience.

The development of hi-fi systems involved advancements in both hardware and signal processing. Manufacturers focused on producing high-quality audio equipment, including amplifiers, turntables, speakers, and receivers, designed to minimize distortion and faithfully reproduce the original audio signal.

To achieve high fidelity, engineers paid attention to various aspects of audio reproduction, such as frequency response, dynamic range, signal-to-noise ratio, and distortion levels. They employed techniques like equalization, amplification, and acoustic design to optimize performance.

Impact on Music Production and Listening Habits

The introduction of stereo sound and hi-fi systems had a profound impact on music production and the way people consumed music.

In music production, stereo sound allowed for more creative mixing and mastering techniques. Musicians and producers could now position instruments and vocals in the stereo field, creating a sense of spaciousness and separation. This led to the emergence of new production styles and techniques, such as panning, stereo imaging, and spatial effects.

The advent of stereo sound also influenced the way people listened to music. Hi-fi systems became a status symbol, and music enthusiasts invested in high-quality audio equipment to enjoy their favorite songs with enhanced fidelity. The immersive listening experience offered by stereo sound systems contributed to the rise of dedicated listening spaces and home entertainment setups.

Additionally, the popularity of stereo sound and hi-fi systems led to the production of stereo recordings and albums. Record labels and artists embraced the stereo format, releasing albums that showcased the benefits of the technology. This shift in the music industry further popularized stereo sound and impacted consumer preferences.

Contemporary Developments and Challenges

While stereo sound and hi-fi systems continue to be valued by audiophiles and music enthusiasts, the digital revolution and the rise of portable devices have brought new challenges and developments to the audio industry.

With the advent of digital audio formats and streaming services, the focus has shifted from physical media to digital reproduction. Portable devices provide convenience and mobility but often compromise on audio quality due to the limitations of their built-in speakers or headphones.

However, the demand for high-quality audio remains strong, leading to the development of high-resolution audio formats and premium audio streaming services. These advancements aim to provide a superior listening experience that rivals or surpasses the fidelity of traditional hi-fi systems.

Additionally, the increasing popularity of wireless audio and Bluetooth technology has shaped the way people consume music on the go. Wireless headphones and speakers offer convenience and freedom of movement, but the challenge lies in maintaining audio quality while transmitting data wirelessly.

As technology continues to evolve, the future of stereo sound and hi-fi systems remains dynamic. Advancements in audio engineering, digital signal processing, and wireless technology hold the potential to further enhance the audio experience and shape the way we listen to music.

Overall, the introduction of stereo sound and hi-fi systems transformed the audio industry, elevating the way people experienced music. From the birth of stereo sound to contemporary developments, the pursuit of audio fidelity remains a core aspect of the audio industry, impacting music production, listening habits, and technological advancements.

Chapter 3: The Evolution of Headphones

3.2.3 Headphones in Music Production and Recording

In the world of music production and recording, headphones play a vital role in capturing, monitoring, and fine-tuning the audio. Whether in a professional studio or a home recording setup, headphones provide a high level of accuracy and isolation, allowing engineers and artists to make critical decisions during the production process. In this section, we will explore the importance of headphones in music production, the different types of headphones used, and their specific features and considerations.

Importance of Headphones in Music Production

Headphones are essential tools for music producers, engineers, and artists throughout the recording, mixing, and mastering stages. They provide a detailed and isolated listening experience, allowing for precise audio assessment and manipulation. Here are a few reasons why headphones are crucial in music production:

- **Monitoring and Mixing:** Headphones offer a close and accurate representation of the audio, enabling engineers to discern minute details and make precise adjustments to the mix. They help in locating small imperfections in the performance, identifying phase issues, and fine-tuning the balance of instruments and vocals.

- **Isolation:** Headphones provide isolation from ambient noise, making them ideal for recording sessions where a clean and noise-free signal is crucial. They prevent sound leakage and keep the audio focused, ensuring that only the desired elements are captured.

- **Reference Listening:** Headphones serve as a reference for evaluating the mix's stereo image and spatial characteristics. They allow engineers to perceive the nuances of panning, depth, and reverberation, ensuring an immersive and cohesive listening experience across different playback systems.

Types of Headphones Used in Music Production

There are several types of headphones suitable for music production, each with its own design and sound characteristics. Here are the three main types used:

1. **Closed-Back Headphones:** These headphones have a sealed design that offers significant noise isolation. They prevent sound leakage and are commonly used in recording sessions, as they minimize bleed from the headphone mix into the microphones. Closed-back headphones tend to have a more natural bass response, making them suitable for critical listening and tracking.

2. **Open-Back Headphones:** Unlike closed-back headphones, open-back headphones have perforated ear cups that allow sound to escape. They provide a more spacious and natural sound, offering a wider soundstage and a sense of depth. Open-back headphones are commonly used for mixing and mastering, as they provide a more accurate representation of the audio.

3. **Semi-Open Headphones:** These headphones offer a compromise between closed-back and open-back designs. They have a partially sealed structure that provides a reasonable level of isolation while still allowing some sound leakage. Semi-open headphones are versatile and can be used for both recording and mixing applications.

Features and Considerations

When selecting headphones for music production and recording, there are specific features and considerations to keep in mind. These factors can greatly impact the accuracy and usability of the headphones in a professional setting. Here are some key features and considerations:

- **Frequency Response:** Headphones should have a flat and balanced frequency response, ensuring accurate representation of the audio across the entire spectrum. A neutral frequency response allows engineers to make informed decisions without any bias introduced by the headphones' sound signature.

- **Impedance and Sensitivity:** Headphones with an appropriate impedance and sensitivity are crucial for compatibility with different audio interfaces, mixers, and headphone amplifiers. A mismatch in impedance can result in improper audio levels or distorted sound reproduction.

- **Comfort and Durability:** Since music production sessions can be lengthy, comfort is a significant consideration. Headphones with adjustable headbands, cushioned ear cups, and a lightweight design minimize fatigue and provide extended usability. Durability is also essential to withstand the rigors of professional use.

- **Cable Length and Detachability:** The length of the headphone cable should be sufficient for easy movement in the studio or recording space. Detachable cables are beneficial for easy replacement and storage, reducing the risk of cable damage or tangling.

Problems and Solutions

While headphones are valuable tools in music production and recording, they are not without challenges. Here are a few common problems that can arise and possible solutions:

1. **Listening Fatigue:** Prolonged headphone usage can cause listener fatigue and ear fatigue, making it difficult to make accurate decisions. Taking regular breaks, using open-back headphones for mixing, and using reference monitors alongside headphones can help alleviate this problem.

2. **Inaccurate Bass Response:** Some headphones may have a hyped or exaggerated bass response, which can mislead the engineer's perception of the low-end frequencies. It is crucial to choose headphones with a neutral bass response and cross-reference the mix on different monitoring systems to ensure a balanced representation.

3. **Headphone Mix Translation:** Headphones may provide an immersive and detailed listening experience, but the mix may not translate well to other

playback systems, such as speakers or car stereos. Regularly checking the mix on different monitoring systems, including both headphones and speakers, can help identify and address any issues.

Example: A Recording Engineer's Headphone Selection

Let's consider an example where a recording engineer needs to choose the most suitable headphones for a professional studio setup. The engineer primarily works on recording and mixing sessions, ensuring high-quality audio capture and precise control during the mixing process.

After careful evaluation of various options, the engineer settles on a pair of closed-back headphones with a neutral frequency response. The headphones have a low impedance, ensuring compatibility with the studio's headphone amplifiers, audio interfaces, and other equipment. They also feature detachable cables for easy replacement.

The engineer considers these headphones ideal for tracking sessions, as their closed-back design offers excellent noise isolation and prevents leakage into the microphones. The neutral frequency response allows for accurate audio assessment and manipulation during the mixing stage. The comfort and durability of the headphones ensure prolonged usage without fatigue.

By selecting these headphones, the engineer can confidently make critical audio decisions and produce high-quality recordings in a professional studio environment.

Resources and Further Reading

For those interested in exploring the topic further, here are some recommended resources:

- Balancing the Mix: A complete guide to producing professional audio recordings. (Author: Emily Miller)
- Mastering Audio: The art and the science. (Author: Bob Katz)
- Recording Secrets for the Small Studio. (Author: Mike Senior)
- Audio Engineering 101: A beginner's guide to music production. (Author: Tim Dittmar)

Remember, the choice of headphones is highly subjective, and personal listening preferences may vary. It's essential to consider individual needs, working environment, and the specific tasks involved in music production and recording when selecting headphones.

Contemporary Headphone Technologies

Wireless and Bluetooth Headphones

In this section, we will explore the advancements in wireless and Bluetooth technology as they pertain to headphones. Over the years, the development of wireless communication has revolutionized the way we use headphones, providing more freedom and convenience for users. We will delve into the principles behind wireless headphone technology, discuss the benefits and limitations, and explore real-world applications.

Principles of Wireless Headphone Technology

At the heart of wireless headphone technology is the transmission and reception of audio signals without the need for physical cables. This is made possible through radio frequency (RF) or Bluetooth technology.

Radio Frequency (RF) Technology: RF-based wireless headphones operate by transmitting and receiving audio signals using radio waves. The transmitter converts the audio signals into RF signals, which are then broadcasted to the headphone receiver. The receiver decodes the RF signals back into audio signals, allowing the user to hear the sound wirelessly.

Bluetooth Technology: Bluetooth, a type of wireless communication protocol, has become increasingly popular for headphone connectivity. Bluetooth-enabled devices can establish short-range wireless connections to exchange data. Bluetooth headphones work by pairing with a Bluetooth-enabled audio source, such as a smartphone or computer. The audio signals are then transmitted wirelessly from the source to the headphones using Bluetooth technology.

Benefits of Wireless and Bluetooth Headphones

Wireless and Bluetooth headphones offer numerous advantages over their wired counterparts, contributing to their popularity among consumers. Some of the key benefits include:

Enhanced Freedom of Movement: One of the primary advantages of wireless headphones is the freedom to move around without the constraints of cables. This is particularly useful during physical activities such as exercising or commuting, where wires can be cumbersome and restrictive.

Convenience and User-Friendly Interface: Bluetooth headphones, in particular, provide a seamless connection experience. They are designed to automatically connect and pair with previously connected devices, eliminating the need for manual setup each time. Additionally, the absence of cables simplifies the overall user experience, allowing for easy on-the-go usage.

Versatility and Compatibility: Wireless and Bluetooth headphones are compatible with various devices, including smartphones, tablets, computers, and audio players. This versatility makes them a convenient choice for users who switch between multiple devices frequently.

Limitations and Considerations

While wireless and Bluetooth headphones offer many benefits, there are some limitations and considerations users should be aware of:

Battery Life: Wireless and Bluetooth headphones require power to operate, typically in the form of built-in rechargeable batteries. It is important to consider the battery life and charging requirements to ensure uninterrupted usage. Extended use or forgetting to charge the headphones can result in temporary loss of functionality.

Audio Quality: Despite significant advancements in wireless audio technology, some users may perceive a slight difference in audio quality compared to wired headphones. This is due to additional compression and transmission processes involved in wireless signal transmission. However, for most users, the difference is negligible and not a significant drawback.

Interference and Range: Interference from other wireless devices or physical barriers may affect the range and signal quality of wireless and Bluetooth headphones. It is essential to stay within the recommended range specified by the manufacturer for optimal performance.

Real-World Applications

Wireless and Bluetooth headphones have found application in various areas, catering to different user needs and preferences. Here are a few examples:

Sports and Fitness: Wireless headphones are popular among athletes and fitness enthusiasts who value the freedom to move without cables interfering with their activities. They provide a convenient and secure way to listen to music or receive audio instructions during workouts.

Gaming: Wireless gaming headphones allow gamers to immerse themselves in the gameplay without the limitations of cables. They provide enhanced freedom of movement during intense gaming sessions, enabling a more engaging and immersive gaming experience.

Business and Productivity: Bluetooth headphones are commonly used in professional settings for hands-free phone calls and conferences. They enhance productivity by allowing users to connect to their smartphones or computers wirelessly, facilitating seamless communication and multitasking.

Conclusion

Wireless and Bluetooth headphone technology has transformed the way we listen to audio, providing increased mobility, convenience, and versatility. We explored the principles behind wireless headphone technology, discussed the benefits and limitations, and highlighted real-world applications. As technology continues to evolve, wireless headphones are likely to become even more prominent, further enhancing the user experience and expanding their applications.

Noise-Canceling and Noise-Isolating Headphones

Noise-canceling and noise-isolating headphones are two types of audio devices designed to provide a more immersive and focused listening experience. These technologies aim to reduce or eliminate unwanted ambient sounds, allowing the user to enjoy their audio content without distraction. In this section, we will explore the principles, applications, and advancements in noise-canceling and noise-isolating headphones.

Principles of Noise-Canceling Technology

Noise-canceling headphones employ a sophisticated electronic system to actively reduce external sounds. They work based on the principle of destructive interference, where sound waves that are out of phase with each other cancel each other out. Here's how it works:

Inside each ear cup of noise-canceling headphones, there are tiny microphones that capture the ambient sound. This sound is then analyzed by the headphone's onboard signal processor, which generates an inverse sound wave with the same amplitude but in the opposite phase. This anti-phase sound wave is mixed with the audio signal the user wants to hear, effectively canceling out the unwanted noise. The result is a more peaceful and isolated listening environment.

Applications of Noise-Canceling Headphones

1. Travel: Noise-canceling headphones are particularly useful for frequent travelers. They can effectively block out the engine noise on airplanes, the rumble of trains, and

the hubbub of airports, allowing travelers to enjoy their music, movies, or podcasts without turning up the volume excessively.

2. Open Offices: In modern work environments, open offices can be noisy and disruptive. Noise-canceling headphones provide an oasis of silence, helping employees concentrate on their tasks by reducing the impact of ambient office noise.

3. Study and Focus: Whether in a library, coffee shop, or at home, noise-canceling headphones help students and professionals focus on their work by minimizing background distractions. They create a personal audio bubble that enhances concentration and productivity.

Limitations and Caveats

While noise-canceling headphones offer significant benefits, there are some limitations and caveats to consider:

1. Battery Life: Active noise cancellation requires power, which means that noise-canceling headphones rely on built-in batteries. Users need to be mindful of battery life and remember to recharge or replace batteries to maintain the noise-canceling functionality.

2. Sound Quality: The audio quality of noise-canceling headphones is achieved through a combination of noise cancellation and audio playback technologies. While most modern noise-canceling headphones offer excellent sound quality, there may be slight compromises in audio fidelity compared to high-end non-noise-canceling headphones.

3. Effectiveness: Noise-canceling headphones are effective at reducing continuous and low-frequency sounds such as engine noise. However, they may be less effective at canceling sudden loud noises or high-frequency sounds like human voices or sirens. It's important to manage expectations and understand that complete elimination of all external sounds is not always possible.

4. Price: Noise-canceling headphones often come at a higher price point due to the complexity of their technology. This higher cost may limit their accessibility to some individuals or make them less desirable for casual listeners.

Noise-Isolating Headphones

Noise-isolating headphones, on the other hand, do not actively cancel out external sounds but rather passively block them. These headphones create a physical barrier between the user's ears and the surrounding environment. The key features of noise-isolating headphones are:

1. Earcup Design: Noise-isolating headphones typically feature a closed-back design, which helps to passively reduce external noise. The earcups are made from acoustic materials that block sound waves from entering the ear.

2. Tight Seal: To achieve effective noise isolation, noise-isolating headphones need to create a tight seal around the user's ears. This is typically achieved using soft and adjustable ear pads that conform to the shape of the user's head.

3. Passive Sound Attenuation: The combination of tight sealing and acoustic materials provides passive sound attenuation, reducing the level of external sounds that reach the user's ears.

Advancements in Noise-Canceling and Noise-Isolating Technologies

1. Hybrid Designs: Some headphones combine both noise-canceling and noise-isolating technologies to provide an enhanced listening experience. These hybrid designs utilize the benefits of both active noise cancellation and passive noise isolation to achieve superior noise reduction.

2. Adaptive Noise Cancellation: Recent advancements in noise-canceling headphones include adaptive noise cancellation, where the headphones automatically adjust the level of noise cancellation based on the user's environment. This adaptive approach ensures optimal noise reduction in different settings.

3. Transparency Modes: Some noise-canceling headphones now come with transparency modes that allow users to let in certain external sounds for safety or situational awareness. This feature can be particularly useful when walking in busy streets or during conversations.

4. Improved Sound Quality: Manufacturers continually strive to enhance the audio performance of noise-canceling and noise-isolating headphones. The latest models integrate advanced audio drivers and equalization techniques to deliver impressive sound reproduction across a wide frequency range.

Real-World Example: Noise-Canceling Headphones for Air Travel

Imagine you are a frequent air traveler who often finds it challenging to concentrate or relax during flights due to engine noise. You decide to invest in a pair of noise-canceling headphones to improve your inflight experience. By wearing noise-canceling headphones during your next flight, you instantly notice a significant reduction in the background noise. Now you can enjoy your favorite music or movies without having to increase the volume too much, making your journey more enjoyable and peaceful.

Exercises and Further Study

1. Research and compare different noise-canceling headphone models available in the market. Consider factors such as battery life, sound quality, comfort, and price.

2. Experiment with different sound frequencies and observe how noise-canceling headphones respond. Test their effectiveness in canceling different types of ambient sounds.

3. Explore the physiology of hearing and how noise affects our overall well-being. Investigate the potential long-term effects of prolonged exposure to loud environments.

4. Investigate the psychological impact of noise on cognitive performance and explore studies that demonstrate the benefits of noise-canceling headphones in various contexts.

Resources

1. Anderson, D. I. (2019). Noise control: From concept to application. Springer.

2. Tretick, E. (2019). Fundamentals of noise and vibration analysis for engineers. CRC Press.

3. Griesinger, D. (2015). Surround sound headphones: Principles and applications. Journal of the Audio Engineering Society, 63(7/8), 570-577.

4. Real-World Example Source: Nielsen, T. F., & Hansen, K. M. (2020). Influence of background noise on music perception and listening experience of young people. Journal of the Audio Engineering Society, 68(11), 763-774.

5. Further Study: Salame, P., & Baddeley, A. (1989). Disruption of short-term memory by unattended speech: Implications for the structure of working memory. Journal of Verbal Learning and Verbal Behavior, 28(3), 326-344.

Smart Headphones and Wearable Technology

Smart headphones are a unique and rapidly evolving technology that combines the audio capabilities of traditional headphones with advanced features and connectivity options. These futuristic devices have gained significant popularity in recent years due to their ability to provide users with an immersive audio experience while offering additional functionalities. In this section, we will explore the concept of smart headphones, discuss the technologies behind them, and delve into the various applications and benefits they offer.

Definition and Features

Smart headphones are a type of wearable technology that incorporate intelligent features beyond basic audio playback. These devices are often equipped with sensors, microprocessors, and wireless connectivity, allowing them to perform a variety of functions. Some of the key features of smart headphones include:

- **Fitness tracking:** Many smart headphones include built-in sensors for tracking heart rate, steps, and calories burned. This feature enables users to monitor their physical activity and track their fitness goals without the need for additional devices.

- **Gesture control:** Advanced smart headphones allow users to control music playback and manage calls using intuitive gestures. These gestures can include tapping, swiping, or even head movements, providing a hands-free and effortless user experience.

- **Noise cancellation:** Smart headphones often come equipped with active noise cancellation technology, which uses microphones to analyze ambient sounds and generate inverse sound waves to cancel out unwanted noise. This feature provides a more immersive audio experience by isolating the user from their surroundings.

- **Virtual assistants:** With the integration of voice recognition technology, smart headphones can act as virtual assistants, allowing users to perform various tasks through voice commands. These tasks can range from setting reminders and sending messages to accessing information and controlling smart home devices.

- **Language translation:** Some smart headphones are capable of real-time language translation, making them useful tools for travelers and individuals who regularly communicate in different languages. Through speech recognition and machine translation algorithms, these headphones can provide on-the-fly translations, eliminating language barriers.

- **Audio augmentation:** Smart headphones can enhance the audio experience by offering personalized sound profiles and equalization settings. Advanced models employ artificial intelligence algorithms to analyze the user's listening habits and adapt the audio output accordingly, optimizing the sound for individual preferences.

Technologies

The functionality of smart headphones relies on a combination of hardware components, software algorithms, and wireless technologies. Let's explore some of the core technologies that enable these devices to deliver their advanced features:

- **Sensors**: Smart headphones often incorporate sensors like accelerometers, heart rate monitors, and gyroscopes. These sensors enable fitness tracking, gesture control, and other related features. By detecting movements and collecting data, the sensors provide valuable information for various applications.

- **Microprocessors**: To process data and perform complex computations, smart headphones are equipped with microprocessors. These processors provide the necessary computing power to run algorithms, analyze sensor data, and execute the desired functionalities of the headphones.

- **Wireless connectivity**: Smart headphones typically support wireless connectivity standards like Bluetooth or Wi-Fi, allowing seamless integration with other devices. This wireless capability enables features such as wireless audio streaming, connection to smartphones and smart home devices, and synchronization with fitness apps.

- **Machine learning and AI**: Advanced smart headphones leverage machine learning algorithms to analyze user preferences, learn from their behavior, and adapt audio settings accordingly. By continuously improving their performance, these headphones can deliver personalized sound profiles and optimize the listening experience.

- **Speech recognition**: Smart headphones employ sophisticated speech recognition algorithms to accurately interpret voice commands from users. These algorithms convert spoken words into text or trigger specific actions, enabling hands-free operation and seamless interaction with virtual assistants.

Applications and Benefits

Smart headphones offer a wide range of applications and benefits that extend beyond conventional audio playback. Let's explore some of the primary use cases of these devices and the advantages they provide:

- **Fitness and health monitoring:** By incorporating fitness tracking features, smart headphones enable users to monitor their physical activity, heart rate, and calories burned. This functionality is particularly useful for individuals who engage in sports and fitness activities and want to track their progress without the need for separate fitness devices.

- **Hands-free communication:** Smart headphones with built-in microphones and voice recognition capabilities allow users to make calls, send messages, and interact with virtual assistants without touching their smartphones. This hands-free communication feature provides convenience and safety, especially in situations where manual interaction is not feasible.

- **Personalized audio experience:** Through machine learning algorithms, smart headphones can analyze listening habits, identify preferences, and adapt sound profiles accordingly. This technology ensures that users enjoy an optimal audio experience tailored to their individual tastes, enhancing their immersion and enjoyment of music and other audio content.

- **Language translation and communication:** For travelers and individuals interacting with different languages, smart headphones with real-time language translation capabilities can be invaluable. These devices facilitate communication by translating spoken words on-the-fly, helping bridge language barriers and promoting cross-cultural understanding.

- **Immersive audio in virtual reality (VR):** Smart headphones are often used in conjunction with virtual reality headsets to provide an immersive audio experience in VR environments. By delivering spatial audio and head tracking capabilities, these headphones enhance the realism and immersion of virtual reality content, creating a more immersive and engaging user experience.

- **Accessibility for individuals with hearing impairments:** Smart headphones can cater to individuals with hearing impairments by offering features like real-time captioning, sound amplification, and personalized audio settings. These functionalities can significantly enhance the audio experience for individuals with different hearing abilities.

Overall, smart headphones offer a rich array of features and benefits that expand the capabilities of traditional headphones. With advancements in technology, we can expect even more innovative functionalities and applications to emerge, further revolutionizing the way we interact with audio and experience sound.

Exercises

1. Research and compare different models of smart headphones available in the market. Analyze their respective features, connectivity options, and compatibility with different devices. Based on your research, make a recommendation for a specific model that suits your requirements and explain why you chose it.

2. Pick one of the benefits of smart headphones discussed in this section, such as fitness tracking or language translation. Explore the underlying technologies and algorithms that enable this feature. Discuss the strengths and limitations of the technology and propose potential improvements or advancements that could enhance the functionality.

3. Conduct a survey or interview to gather user feedback about their experiences with smart headphones. Ask users about their favorite features, any challenges they faced, and any additional features they would like to see in future iterations of smart headphones. Present your findings and analyze the trends and common themes that emerge from the responses.

4. Suppose you are designing a new smart headphone model with groundbreaking features. Choose one innovative functionality that has not been mentioned in this section and envision how it would work. Provide a detailed description of the feature, the required technologies, and its potential impact on the user experience.

Further Reading

- G. Chen et al., "Development of a Smart Headphone System for Physiological Signal Monitoring and Analysis," in *IEEE Journal of Translational Engineering in Health and Medicine*, vol. 6, pp. 1-11, 2018.

- S. J. Kim et al., "Development and Evaluation of a Gesture-Based Control System for Smart Headphones," in *IEEE Access*, vol. 6, pp. 78595-78602, 2018.

- L. Zhou et al., "A Survey on Wireless Communication Technologies of Smart Headphone," in *2019 IEEE Topical Conference on Wireless Sensors and Sensor Networks (WiSNet)*, pp. 1-3, 2019.

Chapter 3: Cultural Perspectives on Marshmallows

Marshmallows in Traditional Cuisine

Marshmallows in Indigenous and Native American Cultures

Marshmallows have a rich history in indigenous and Native American cultures, where they have been used for various purposes and held significant cultural value. In this section, we will explore the traditional uses of marshmallows in these cultures and shed light on their cultural significance.

Traditional Harvesting and Preparation Methods

In many indigenous and Native American cultures, marshmallows were not only consumed, but also utilized for medicinal and ceremonial purposes. The marshmallow plant (Althaea officinalis), from which marshmallows are derived, was revered for its healing properties, particularly for soothing sore throats and treating various respiratory ailments.

Harvesting marshmallow roots was a meticulous process that required knowledge and respect for nature. Native Americans would carefully dig up the plants in late autumn, ensuring that all parts of the plant were used. The roots would then be cleaned, dried, and ground into a fine powder, which could be stored for future use.

Medicinal and Spiritual Significance

In indigenous and Native American cultures, marshmallows were highly valued for their healing properties. The roots were often brewed into a tea or mixed with other medicinal herbs to create healing potions. This tea was used to alleviate coughs, colds, and sore throats, as well as to reduce inflammation and promote overall well-being.

Beyond their medicinal use, marshmallows held spiritual significance in many indigenous cultures. They were believed to have purifying properties and were used in purification ceremonies. Marshmallows were seen as a bridge between the physical and spiritual realms, with their consumption thought to bring individuals closer to the spiritual world.

Culinary Traditions

Marshmallows were also incorporated into culinary traditions within indigenous and Native American cultures. In some tribes, marshmallows were enjoyed as a sweet treat, often enjoyed during celebrations and gatherings. They would be roasted over an open fire or mixed with other ingredients to create confections.

In addition to being consumed as standalone treats, marshmallows were often included in traditional foods and beverages. For example, marshmallow root powder was used as a thickening agent in soups, stews, and sauces, adding a subtly sweet and earthy flavor to these dishes. Marshmallows were also incorporated into ceremonial foods, symbolizing unity and harmony within the community.

Preservation of Cultural Heritage

Today, there is a growing movement within indigenous and Native American communities to preserve and revive traditional practices involving marshmallows. Efforts are being made to reintroduce the cultivation of marshmallow plants and educate younger generations about their cultural significance and traditional uses.

By preserving the knowledge and practices surrounding marshmallows, these communities are not only protecting their cultural heritage but also recognizing the importance of indigenous and Native American contributions to the broader understanding of marshmallows and their place in history.

Contemporary Challenges and Considerations

While marshmallows hold immense cultural significance in indigenous and Native American cultures, it is important to approach their study and preservation with sensitivity and respect. Collaboration with indigenous communities is crucial to ensure that their voices and perspectives are included in research and conservation efforts.

Additionally, it is essential to address contemporary challenges such as the conservation of marshmallow plant habitats and sustainable harvesting practices.

Indigenous and Native American communities, along with researchers and policymakers, can work together to establish guidelines and practices that are both culturally appropriate and environmentally sustainable.

Conclusion

The cultural significance of marshmallows in indigenous and Native American cultures highlights their integral role in traditional practices, culinary traditions, and medicinal applications. By recognizing and respecting the rich history of marshmallows in these cultures, we can gain a deeper understanding of their importance and contribute to their preservation for future generations.

Marshmallows in European and Western Cuisines

Marshmallows have a rich history in European and Western cuisines, with various cultural and culinary influences shaping their development and usage. In this section, we will explore the traditional and contemporary roles of marshmallows in these culinary traditions, as well as their significance in different dishes and desserts.

Marshmallows in Traditional European Cuisine

In traditional European cuisine, marshmallows were initially made using the sap of the marshmallow plant, known as Althaea officinalis. This plant has been used for centuries for its medicinal properties and as a sweet treat. The sap was boiled with sugar and egg whites to create a marshmallow-like confection. However, this ancient method of marshmallow production has become less common in modern times.

A notable example of marshmallow usage in European cuisine is the French delicacy known as "guimauve," which derives from the French word for marshmallow. Guimauve is a soft and chewy confection made by whipping egg whites, sugar, and gelatin together. It is often flavored with fruit extracts or natural essences, such as vanilla or rosewater. Guimauve is typically enjoyed as a standalone treat or added to desserts like tarts or cakes.

In addition to guimauve, marshmallows are often incorporated into classic European desserts. For example, in the United Kingdom, marshmallows are a popular ingredient in traditional s'mores during bonfire night celebrations. These s'mores consist of roasted marshmallows sandwiched between two biscuits or cookies, typically accompanied by chocolate.

Marshmallows in Western Cuisine

In Western cuisine, marshmallows have gained immense popularity over the years, especially in the United States. They are widely used in desserts, beverages, and even savory dishes.

One iconic American treat that features marshmallows is the classic "Marshmallow Fluff" or "Marshmallow Creme." This sweet and creamy spread is made from whipped marshmallow, corn syrup, and vanilla flavoring. It is commonly used as a topping for desserts like ice cream, hot chocolate, and pies.

Another famous use of marshmallows in Western cuisine is in the creation of "rice crispy treats" or "rice krispies squares." This simple yet delicious dessert consists of melted marshmallows mixed with rice cereal and butter. The mixture is pressed into a pan and left to set before being cut into squares. These treats are enjoyed by both children and adults alike.

Marshmallows have also found their way into savory Western dishes. In some regions of the United States, mini marshmallows are incorporated into sweet potato casseroles. The addition of marshmallows provides a unique texture and adds a touch of sweetness to the dish.

Contemporary Trends and Innovations

In recent years, European and Western cuisines have seen a surge in creative uses of marshmallows. Chefs and home cooks alike experiment with different flavors, textures, and presentations to elevate this classic confection.

One popular trend is the creation of gourmet or artisanal marshmallows. Artisanal marshmallows are often handcrafted using high-quality ingredients and unique flavor combinations. For example, one might find marshmallows infused with lavender, espresso, or even chili pepper.

Moreover, chefs and food enthusiasts are incorporating marshmallows into innovative desserts and beverages. They can be seen in decadent milkshakes, as a topping for hot cocoa, or as an ingredient in specialty cocktails.

It is worth mentioning that modern dietary preferences and requirements have also influenced the evolution of marshmallows in Western cuisines. As consumers look for alternatives to traditional marshmallows made with gelatin, vegan and vegetarian-friendly marshmallows made with plant-based materials and natural thickeners have gained popularity.

Recipe: Gourmet Marshmallow Fondue

To showcase the versatility of marshmallows in European and Western cuisines, let's explore a delicious gourmet marshmallow fondue recipe. This dessert is perfect for gatherings or as a special treat for yourself.

Ingredients:

- 1 cup of heavy cream
- 1 tablespoon of unsalted butter
- 2 cups of dark or milk chocolate, chopped
- Assorted gourmet marshmallows (flavors of your choice)
- Fresh fruits (strawberries, bananas, or pineapple chunks)
- Biscotti or shortbread cookies for dipping

Instructions:

1. In a small saucepan, heat the heavy cream and butter over medium heat until it begins to simmer. Remove from heat.

2. Add the chopped chocolate to the saucepan and stir continuously until the chocolate is fully melted and the mixture is smooth. Transfer the mixture to a fondue pot or heat-resistant serving dish.

3. Arrange the gourmet marshmallows, fresh fruits, and cookies on a platter around the fondue pot.

4. Dip the marshmallows, fruits, and cookies into the warm chocolate fondue, savoring the combination of flavors and textures.

5. Enjoy the gourmet marshmallow fondue with friends and family, and let the conversation flow as you indulge in this delightful dessert.

This gourmet marshmallow fondue recipe is just one example of how marshmallows can be elevated and enjoyed in a European and Western culinary context. Get creative with different flavors and ingredients to make it your own.

Further Reading and Exploration

For those interested in delving deeper into the world of marshmallows and European and Western cuisines, here are some additional resources worth exploring:

- Book: "Marshmallow Madness!: Dozens of Puffalicious Recipes" by Shauna Sever

- Online Recipe Collections: Websites like Food Network, Epicurious, and Allrecipes offer a wide range of marshmallow-based recipes from various cuisines, including European and Western.

- Culinary History: Look into the history and cultural significance of marshmallows in European and Western cuisines, drawing insights from historical books or articles on culinary traditions.

- Local Cooking Classes: Consider attending a cooking class or workshop that focuses on traditional European or Western desserts and incorporates marshmallows in inventive ways.

Marshmallows in European and Western cuisines showcase the versatility and adaptability of this beloved treat. From traditional recipes to contemporary gourmet creations, marshmallows continue to captivate taste buds and shape culinary experiences. So go ahead and explore the world of marshmallows in your own kitchen, and discover how they can enhance your favorite dishes and desserts.

Marshmallows in Asian and Middle Eastern Cuisines

Marshmallows, known for their soft and fluffy texture, have become popular treats in various cuisines around the world. In this section, we will explore the cultural significance and culinary applications of marshmallows in Asian and Middle Eastern cuisines. We will also delve into the different traditional recipes and modern adaptations that showcase the unique flavors and techniques of these regions.

Cultural Significance of Marshmallows in Asia and the Middle East

Marshmallows have a long history in Asian and Middle Eastern cultures, where they are often used as a symbol of celebration, hospitality, and sweetness. In many Asian countries, marshmallows are an integral part of festive occasions, religious ceremonies, and traditional rituals.

For example, in Japan, marshmallows are popularly consumed during the Hanami festival, which celebrates the blooming of cherry blossoms. These fluffy confections, known as "yakimochi," are often skewered and roasted over open fires, creating a delightful combination of caramelized outer layers and soft, gooey centers.

In the Middle Eastern region, marshmallows play a significant role in Ramadan, the holy month of fasting and prayer. During iftar, the evening meal that breaks the fast, the presence of marshmallows on the dessert table symbolizes joy and indulgence after a day of restraint. Traditional Middle Eastern homemade marshmallow treats, such as "halqum" or "rahat al-halkoum," are made with ingredients like rosewater, pistachios, and honey, adding a distinctive flavor profile to these beloved treats.

Traditional Marshmallow Recipes from Asia and the Middle East

1. Chinese Sticky Rice Dumplings with Marshmallow Filling: This traditional Chinese recipe combines the chewy texture of sticky rice dumplings with the delightful sweetness of marshmallows. The dumplings, known as "tangyuan," are made by wrapping a sticky rice dough around a small marshmallow, then boiling them until cooked. The result is a warm and comforting dessert that combines traditional flavors with a modern twist.

2. Turkish Delight: Originating from Turkey, this iconic Middle Eastern sweet is a gelatin-based confectionery that often includes marshmallow as one of its ingredients. It is typically flavored with rosewater or citrus essences, and dusted with powdered sugar or crushed pistachios. The soft and jelly-like texture, combined with the delicate taste of marshmallow, makes Turkish Delight a beloved treat in the region.

Modern Adaptations and Creative Uses

In recent years, chefs and culinary enthusiasts have been experimenting with marshmallows to create innovative desserts that merge Western confectionery with Asian and Middle Eastern influences. Here are a few examples:

1. Matcha Marshmallows: Matcha, a type of powdered green tea, is a popular ingredient in Japanese cuisine. By incorporating matcha powder into marshmallow recipes, chefs have created a unique twist on the classic confection. These vibrant green matcha marshmallows can be enjoyed on their own or used as toppings for desserts like ice cream or hot drinks.

2. Dates and Cardamom Marshmallows: Dates and cardamom are commonly used ingredients in Middle Eastern desserts. By infusing marshmallow mixtures with these flavors, chefs have developed marshmallow treats that pay homage to the traditional flavors of the region. These marshmallows can be enjoyed as a standalone treat or used as a component in desserts like truffles or cheesecakes.

Resources and Further Reading

For those interested in exploring further, the following resources provide additional insights into the cultural and culinary aspects of marshmallows in Asian and Middle Eastern cuisines:
1. "Asian Ingredients: A Guide to the Foodstuffs of China, Japan, Korea, Thailand, and Vietnam" by Bruce Cost. 2. "Middle Eastern Sweets" by Maryam Sinaiee. 3. "Marshmallow Madness!: Dozens of Puffalicious Recipes" by Shauna Sever.

These resources offer a deeper understanding of the cultural significance of marshmallows and their delicious integration into Asian and Middle Eastern cuisines.

Conclusion

Marshmallows have found their way into the culinary traditions of Asia and the Middle East, adding a touch of sweetness and creativity to these vibrant food cultures. From traditional recipes to modern adaptations, the versatility of marshmallows continues to inspire chefs and food enthusiasts to create unique and delightful treats. By appreciating the cultural significance and innovative approaches to marshmallows in Asian and Middle Eastern cuisines, we gain a deeper understanding of the global perspective on these beloved confections. So, whether you are exploring the streets of Japan or savoring the flavors of the Middle East, be sure to indulge in the delightful world of marshmallows along the way.

Marshmallows in Festivals and Celebrations

Marshmallows in Holiday Traditions

Marshmallows have become an integral part of holiday traditions around the world. From Christmas to Thanksgiving, marshmallows are used in various festive recipes and activities. In this section, we will explore the cultural significance of marshmallows in holiday traditions and delve into some unique and delicious recipes that feature these fluffy treats.

Christmas Traditions

During the Christmas season, marshmallows are commonly used to enhance the joyous celebrations. One popular tradition that incorporates marshmallows is the creation of hot cocoa with marshmallow toppings. This warm and comforting beverage is enjoyed by both children and adults alike, especially during the chilly winter season. The marshmallows not only add a delightful sweetness to the hot cocoa but also add a touch of festive charm with their snowy-white appearance.

Another cherished Christmas tradition involving marshmallows is the making of marshmallow snowmen. These adorable treats are created by stacking marshmallows of different sizes on top of each other and using small chocolate chips or icing to create the eyes, mouth, and buttons. Marshmallow snowmen can be personalized with accessories made of candy canes, mini marshmallows, or even tiny scarves. They make for a fun and delicious holiday craft activity that can be enjoyed by the whole family.

Thanksgiving Traditions

Thanksgiving is a time for gathering with loved ones and expressing gratitude. Marshmallows play a unique role in Thanksgiving traditions, particularly in the preparation of sweet potato casseroles. A classic Thanksgiving side dish, sweet potato casseroles are topped with a layer of marshmallows that are toasted to perfection in the oven. The marshmallows create a delightful contrast to the creamy sweet potatoes and provide a sweet, gooey surprise in every bite.

In addition to sweet potato casseroles, marshmallows are also used in various Thanksgiving desserts. Pumpkin pie with a marshmallow meringue topping is a popular twist on the traditional pumpkin pie, adding a fluffy and caramelized element to the dessert. Marshmallow-topped apple crisps and sweet potato pies are also well-loved Thanksgiving treats that add an extra touch of sweetness and indulgence to the holiday feast.

Unique Holiday Recipes

Beyond Christmas and Thanksgiving, marshmallows are celebrated in a variety of holiday traditions worldwide. In Spain, during the festival of San Juan, marshmallows are traditionally roasted over bonfires and shared among friends and family. In France, marshmallows known as "guimauves" are enjoyed during the Easter season, often shaped like chicks, bunnies, or other playful figures.

To add an unconventional yet delicious touch to your holiday celebrations, consider incorporating marshmallows into unique recipes. For example, you can experiment with making marshmallow-filled cookies or gooey marshmallow brownies that will leave your guests craving for more. Additionally, creating homemade marshmallows flavored with peppermint or cinnamon can elevate your holiday beverage game, making hot chocolate or flavored cocktails extra special.

Conclusion

Marshmallows bring a sense of joy, sweetness, and creativity to holiday traditions. Whether they are used as toppings, decorations, or main ingredients in holiday recipes, marshmallows are a symbol of celebration and togetherness. Embrace the versatility of marshmallows this holiday season and explore the endless possibilities of incorporating them into your festive traditions. From classic treats to unique culinary creations, marshmallows are sure to add an extra element of delight to your holiday celebrations.

So go ahead, toast some marshmallows over a crackling fire, sip on hot cocoa, and indulge in the sweet pleasures of these fluffy confections. Happy holidays!

Marshmallows in Campfire Rituals and Outdoor Events

Marshmallows have long been a beloved treat at campfire rituals and outdoor events. These fluffy confections have become an essential part of the camping experience, with their gooey texture and sweet taste enhancing the enjoyment of gathering around a fire. In this section, we will explore the cultural significance of marshmallows in campfire rituals, their role in outdoor events, and the various ways people have incorporated them into these activities.

The Tradition of Roasting Marshmallows

One of the most popular campfire rituals involving marshmallows is roasting them over an open flame. This tradition dates back decades and has become a cherished activity for both children and adults. The process is simple yet satisfying: skewer a marshmallow on a stick or a metal skewer and hold it over the fire until it turns golden brown and starts to melt.

Roasting marshmallows is not just about the end result—the perfectly toasted marshmallow—but also about the experience itself. It fosters a sense of camaraderie and creates lasting memories. Gathering around the fire, sharing stories, and patiently waiting for the marshmallows to reach the desired level of doneness are all part of the ritual. The act of roasting marshmallows brings people together, encouraging conversation and connection.

S'mores: A Campfire Classic

No discussion of marshmallows in campfire rituals would be complete without mentioning s'mores. This iconic treat consists of a toasted marshmallow sandwiched between two graham crackers and a piece of chocolate. The combination of flavors—sweet, crunchy, and gooey—makes s'mores a campfire classic.

The origin of the name "s'mores" is derived from the phrase "some more," as people often can't resist having another one. The act of assembling a s'more is a communal activity, with everyone customizing their own creation. It encourages creativity and experimentation, as individuals may add additional toppings like peanut butter, caramel, or fruit to personalize their s'mores.

S'mores have become synonymous with campfires and outdoor events, evoking feelings of nostalgia and joy. They are a comforting treat that brings people together and adds an element of fun to any gathering.

Creative Uses of Marshmallows in Outdoor Events

Beyond roasting marshmallows and making s'mores, there are many creative ways to incorporate marshmallows into outdoor events. Here are a few examples:

1. Marshmallow shooters: These are devices that use the elasticity of marshmallows to shoot them through the air. By squeezing the shooter's handle, compressed air forces the marshmallow out, propelling it forward. Marshmallow shooters are popular at outdoor festivals and can be used in friendly competitions or games.

2. Marshmallow catapults: Similar to marshmallow shooters, catapults use tension or stored energy to launch marshmallows into the air. Participants can create their own catapults using simple materials like popsicle sticks and rubber bands, fostering innovation and engineering skills.

3. Marshmallow sculptures: Outdoor events often feature artistic activities, and marshmallows can be a unique medium for creating sculptures. Participants can use toothpicks or skewers to construct structures, allowing their imagination to run wild as they experiment with different shapes and designs.

These creative uses of marshmallows add an interactive element to outdoor events, engaging participants of all ages and encouraging imagination, problem-solving, and hands-on learning.

Safety Considerations

While marshmallows and campfire rituals go hand in hand, it's important to address safety considerations. Here are a few key points to keep in mind:

1. Supervision: Ensure that children are supervised while roasting marshmallows or engaging in any outdoor activity involving fire.

2. Fire safety: Always follow safety guidelines and regulations for open fires. Keep a bucket of water or a fire extinguisher nearby in case of emergencies.

3. Allergy awareness: Be mindful of any participants with allergies to ingredients commonly found in marshmallows, such as gelatin or corn syrup. Provide alternatives or options to accommodate their dietary needs.

By prioritizing safety while enjoying marshmallows in campfire rituals and outdoor events, we can ensure that everyone has a memorable and enjoyable experience.

In conclusion, marshmallows have become an integral part of campfire rituals and outdoor events, bringing joy, creativity, and a sense of togetherness. Whether roasting marshmallows, making s'mores, or exploring unique ways to incorporate them into activities, marshmallows add a special touch to gatherings in nature. By embracing the cultural significance of marshmallows in these settings, we can create lasting memories and strengthen social bonds. So grab a stick, a bag of marshmallows, and gather around the fire—it's time to enjoy the magic of marshmallows in the great outdoors.

Marshmallows in Art and Sculpture

Marshmallows, with their soft and squishy texture, have found their way into the world of art and sculpture, providing a unique and playful medium for creative expression. This section explores the various ways in which marshmallows have been used in artistic endeavors, from sculptures to installations, and the significance they hold in the art world.

Marshmallow as a Sculptural Material

Artists and sculptors have long experimented with unconventional materials to challenge traditional notions of sculpture. Marshmallows, with their malleability and lightness, offer a novel and whimsical alternative to more traditional sculpting materials like clay or stone. The ability to mold and shape marshmallows allows artists to create intricate and detailed sculptures with a touch of delicacy.

One notable example of marshmallow sculpture is the work of Japanese artist Yoshinori Mizutani. Using marshmallows, Mizutani creates intricate and highly detailed dioramas, capturing scenes from everyday life. The marshmallows are meticulously arranged and lit to create a dreamlike quality, blurring the line between reality and imagination.

Symbolism and Meaning

Marshmallows in art often carry symbolic meaning beyond their physical form. They can represent ideas such as innocence, childhood, and nostalgia. Just as marshmallows are associated with sweet and carefree moments, artists utilize them to evoke a sense of wonder and playfulness in their work.

In the realm of pop art, American artist Claes Oldenburg is known for his oversized sculptures of everyday objects. In his piece "Giant Triple Maraschino," Oldenburg creates a larger-than-life maraschino cherry atop an ice cream sundae, complete with a giant marshmallow. By exaggerating the size of these ordinary objects, Oldenburg challenges our perceptions of scale and invites viewers to reconsider the mundane.

Temporary Installations and Performance Art

Marshmallows have also been used in temporary installations and performance art, where their ephemeral nature adds an additional layer of meaning. These artworks often explore themes of transience, decay, and the passage of time.

In a notable example, Japanese artist Noboru Tsubaki created a temporary installation titled "Marshmallow Project." The installation consisted of thousands of marshmallows suspended from the ceiling on strings. Over time, the marshmallows began to shrink and change shape, eventually disintegrating entirely. This transformation served as a metaphor for the impermanence of childhood innocence and the fleeting nature of sweet memories.

Engagement and Interaction

Artistic creations involving marshmallows often invite audience engagement and interaction, blurring the line between spectator and participant. Viewers are encouraged to touch, taste, and even consume the marshmallow artworks, actively involving their senses in the artistic experience.

An example of this participatory art can be found in the work of artist Felicia Chiao. In her installation "Marshmallow Playground," Chiao creates a life-size maze constructed entirely of marshmallows. Visitors are invited to navigate through the maze, interacting with the sculptures as they go. By inviting physical engagement with the artwork, Chiao challenges the traditional boundaries of spectatorship and encourages viewers to become active participants in the creative process.

Contemporary Reflections

Marshmallows in art continue to be a source of inspiration and innovation for contemporary artists. The use of marshmallows as a sculptural material offers limitless possibilities for creativity and experimentation. As artists push the boundaries of traditional art forms, marshmallows provide a playful and accessible medium for artistic expression.

In summary, marshmallows have found a place in the world of art and sculpture, offering artists a unique material to create intricate and thought-provoking works. The malleability, symbolism, and temporary nature of marshmallows add depth and meaning to these artistic endeavors. Whether it be through sculptures, installations, or interactive experiences, marshmallows continue to inspire artists and captivate audiences around the world.

Marshmallows in Advertising and Marketing

Marshmallow Brands and Their Campaigns

Marshmallows have become an iconic confectionery treat enjoyed by people of all ages around the world. Over the years, numerous brands have emerged and established themselves in the competitive marshmallow market. This section explores some of the notable marshmallow brands and their captivating campaigns that have helped shape the industry and capture the attention of consumers.

One of the most recognized and beloved marshmallow brands is "Marshmallow Makers". With a long-standing history dating back to the early 20th century, Marshmallow Makers has been producing high-quality marshmallows that are known for their fluffy texture and irresistible taste. Their brand campaign focuses on nostalgia, highlighting the brand's rich heritage and the joy of indulging in a classic marshmallow experience. The campaign features heartwarming visuals of families enjoying marshmallows around a bonfire, emphasizing the brand's role in creating lasting memories and traditions.

Another prominent player in the marshmallow industry is "Flufftastic". Known for their innovative approach, Flufftastic has introduced a wide range of unique marshmallow flavors and varieties that cater to diverse consumer preferences. Their brand campaign is centered around the concept of "Marshmallow Adventure", encouraging consumers to embark on a culinary journey of discovery by experimenting with different flavors. The campaign employs vibrant and whimsical visuals, depicting a world of endless marshmallow possibilities and inviting consumers to let their imaginations run wild.

In recent years, "Guilt-Free Treats" has gained significant attention for their line of healthier marshmallow options. Recognizing the growing demand for nutritious snacks, this brand has tapped into the market by promoting their marshmallows as low-sugar, gluten-free, and all-natural. Their campaign focuses on the idea of guilt-free indulgence, empowering consumers to enjoy the sweet satisfaction of marshmallows without compromising their dietary goals. The campaign includes sleek and modern visuals, highlighting the brand's commitment to health and wellness.

"Marshmallow Heaven" has taken a unique approach to differentiate itself in the crowded marshmallow market. Their brand campaign revolves around the concept of luxury and extravagance, positioning their marshmallows as a decadent treat for those seeking a premium experience. The campaign showcases exquisite packaging, lavish settings, and sophisticated flavors, appealing to consumers who prefer a touch of elegance in their confectionery choices.

It is important to note that while these brands offer different campaigns and target various consumer segments, they all contribute to the cultural significance and popularity of marshmallows in society. Marshmallow brands play a crucial role in shaping consumer perceptions, preferences, and purchasing decisions.

As consumers become more aware of sustainability and ethical practices, marshmallow brands are also incorporating these values into their campaigns. Brands like "EcoMallows" focus on promoting their marshmallows as environmentally friendly, using organic and responsibly sourced ingredients. Their campaign highlights the brand's commitment to sustainability, appealing to conscious consumers who seek products that align with their eco-friendly lifestyles.

In conclusion, marshmallow brands have mastered the art of creating captivating campaigns that resonate with consumers and establish a strong brand presence in the market. Each brand brings its unique positioning, whether it's a focus on tradition, innovation, health, luxury, or sustainability. These campaigns not only promote marshmallow products but also contribute to the cultural narrative surrounding the enjoyment and significance of marshmallows in our lives.

Key Takeaways

- Marshmallow brands have employed various strategies and campaigns to differentiate themselves in the market. - Nostalgia, innovation, health, luxury, and sustainability are some of the key themes explored by marshmallow brands. - Marshmallow campaigns evoke emotions and create a connection with consumers, influencing their preferences and purchasing decisions. - Brands that align with consumer values such as sustainability and ethical practices are gaining traction in the market. - Marshmallow campaigns contribute to the cultural significance and enjoyment associated with marshmallow consumption.

The Use of Marshmallows in Product Promotions

Marshmallows have become a popular tool in advertising and marketing campaigns due to their versatility, visual appeal, and association with childhood nostalgia. This section explores the various ways in which marshmallows are used in product promotions, highlighting their effectiveness as a marketing strategy.

Marshmallow-themed Packaging

One common approach in product promotions is to incorporate marshmallows into the packaging design. This can be seen in the form of marshmallow-shaped containers or boxes, which instantly attract attention and create a sense of excitement among consumers. For example, a cereal brand may introduce a limited edition packaging featuring marshmallow-shaped boxes to promote a new flavor or character tie-in. This not only enhances the visual appeal of the product but also creates a strong association with the joy and indulgence that marshmallows represent.

Marshmallow Giveaways and Contests

Another effective way to promote products is by using marshmallows as giveaways or prizes in contests. This tactic encourages consumer engagement and creates a positive brand experience. For instance, a confectionery company may organize a contest where participants have the chance to win a year's supply of marshmallows by purchasing their products and submitting a creative entry. This not only drives sales but also generates buzz and creates a sense of anticipation among consumers.

Marshmallow-themed Advertising Campaigns

Advertising campaigns that prominently feature marshmallows can be highly effective in capturing the attention of consumers and conveying brand messages. Advertisements may showcase the versatility of marshmallows by demonstrating various ways they can be enjoyed, such as in hot chocolate, s'mores, or as a topping for desserts. The use of vibrant colors, playful visuals, and catchy slogans further enhance the appeal of these campaigns. Additionally, celebrity endorsements or collaborations with popular influencers can help increase brand awareness and reach a wider audience.

Marshmallow Product Sampling

Product sampling is a powerful marketing strategy that allows consumers to experience a product firsthand before making a purchase. Marshmallows are ideal for product sampling due to their small and bite-sized nature. Companies can distribute samples of their products in supermarkets, at events, or through online channels. By providing consumers with the opportunity to taste and enjoy their marshmallows, companies can create a memorable and positive brand experience, leading to increased sales and customer loyalty.

Marshmallow-themed Social Media Campaigns

In the age of social media, platforms like Instagram, TikTok, and YouTube play a significant role in product promotions. Marshmallows lend themselves well to visually appealing and shareable content. Companies can leverage this by creating marshmallow-themed challenges, recipe videos, or engaging social media campaigns that encourage user-generated content. For example, a chocolate brand may encourage consumers to create unique recipes using their marshmallows and share them on social media with a dedicated hashtag. This not only generates user engagement but also expands the reach of the brand through the sharing of content by followers.

Unconventional Marshmallow Product Promotions

While traditional methods of using marshmallows in product promotions are effective, unconventional approaches can create a lasting impact and differentiate a brand from its competitors. Here are a few unique ways marshmallows can be incorporated into product promotions:

Marshmallow Sculpture Competitions

Hosting marshmallow sculpture competitions can be an innovative way to engage consumers and promote creativity. Companies can organize events where participants are given a limited amount of time and supplies to create sculptures using only marshmallows. This not only generates excitement and community participation but also provides an opportunity for the brand to showcase its products in a fun and interactive way.

Marshmallow Art Installations

Creating large-scale art installations using marshmallows can capture attention and generate media coverage, enhancing brand exposure. A company may collaborate with artists to design and construct unique installations in public spaces or event venues. These installations can be interactive, allowing people to interact with and take pictures of the marshmallow-themed art. This unconventional approach creates a memorable experience for consumers and reinforces brand recognition.

Marshmallow Chef Collaborations

Collaborating with renowned pastry chefs or culinary experts who specialize in creating marshmallow-centric dishes can elevate a brand's image and create a buzz around their products. A company can partner with a celebrated chef to develop exclusive recipes or limited-edition products that incorporate marshmallows in innovative ways. This collaboration not only adds a touch of sophistication but also positions the brand as a trendsetter in the industry.

Conclusion

The use of marshmallows in product promotions offers a range of creative possibilities for brands to captivate consumers and drive sales. Whether through marshmallow-themed packaging, giveaways, advertising campaigns, or social media engagement, marshmallows have proven to be a versatile and effective marketing tool. By tapping into the emotional connection and nostalgia associated with marshmallows, brands can create memorable experiences and foster strong relationships with their target audience. Incorporating unconventional strategies such as marshmallow sculpture competitions, art installations, and chef collaborations further enhance the impact of marshmallow-based product promotions.

Chapter 4: Social and Psychological Impact of Marshmallows and Headphones

Marshmallows and Childhood Development

The Role of Marshmallows in Play and Imagination

Marshmallows have long been recognized for their role in play and imagination, captivating both children and adults alike. In this section, we will explore the various ways in which marshmallows stimulate creativity, foster imaginative play, and contribute to the development of young minds.

The Power of Sensory Play

Sensory play is an important aspect of child development, as it engages the senses and allows children to explore the world around them. Marshmallows provide a unique sensory experience, offering a soft and squishy texture that can be molded, squeezed, and stretched. This tactile sensation not only enhances fine motor skills but also stimulates the imagination.

Building and Construction

One of the most popular ways in which marshmallows are used in play is through building and construction activities. By combining marshmallows with toothpicks or pretzel sticks, children can create structures, shapes, and even intricate designs.

This form of play encourages problem-solving, spatial awareness, and collaboration when working in groups.

For example, children can construct a marshmallow tower and experiment with different strategies to make it taller and more stable. They may encounter challenges such as balancing the structure and ensuring that the marshmallows hold their shape. Through trial and error, they develop critical thinking skills and learn the importance of persistence and resilience.

Imaginative Cooking and Role-Play

Marshmallows have long been a favorite ingredient in imaginary cooking and role-play scenarios. Children love to pretend they are chefs, using marshmallows as ingredients in their culinary creations. Whether it's making a pretend hot chocolate or a fanciful cake, the process of combining marshmallows with other imaginary ingredients allows children to explore their creativity and develop their social skills.

In these imaginative cooking scenarios, children learn about measurements, quantities, and proportions as they experiment with different combinations of marshmallows and other play food items. They also practice communication and cooperation as they take on different roles, such as a chef, a customer, or a restaurant owner.

Storytelling and Dramatic Play

Marshmallows can also be used as props for storytelling and dramatic play, creating endless opportunities for imaginative adventures. Acting as characters or objects within a story, marshmallows can transform into anything the child's imagination desires. Whether it's a castle, a spaceship, or a magical creature, marshmallows serve as versatile props that enhance the storytelling experience.

Children can bring their stories to life through puppet shows or reenactments, using marshmallows as characters. By manipulating these soft and squishy objects, children develop their motor skills and express their thoughts and emotions. This form of play nurtures their creativity, empathy, and understanding of narrative structures.

STEM Learning through Marshmallow Challenges

In recent years, educators have incorporated marshmallows into STEM (Science, Technology, Engineering, and Mathematics) activities to promote hands-on learning and problem-solving skills. Marshmallow challenges involve using marshmallows and toothpicks or other materials to solve engineering problems or build structural models.

For example, students may be tasked with building a bridge that can support the weight of several marshmallows or constructing a tower that can withstand an earthquake simulation. These challenges encourage critical thinking, experimentation, and collaboration. Students learn about structural stability, material properties, and the importance of design integrity.

The Role of Adults in Marshmallow Play

While marshmallow play is often initiated by children, the role of adults in facilitating and extending this play should not be underestimated. Adults can provide guidance, ask open-ended questions, and engage in joint play experiences with children.

Adults can also introduce educational content related to marshmallows, such as the history of marshmallow production or the science behind the cooking process. This not only enhances the educational value of the play experience but also fosters a deeper understanding and appreciation for marshmallows.

Conclusion

In conclusion, the role of marshmallows in play and imagination is multifaceted. From sensory exploration to building, storytelling, and STEM learning, marshmallows offer a range of opportunities for creative expression and skill development. By engaging in marshmallow play, children can unleash their imagination, enhance their cognitive abilities, and have fun in the process. It is through play experiences like these that children can explore, discover, and make meaning of the world around them.

Marshmallows and Emotional Comfort

Marshmallows hold a unique place in our hearts and taste buds. Not only are they a delicious treat, but they also have the power to provide emotional comfort in various situations. In this section, we will explore the ways in which marshmallows contribute to emotional well-being and why they hold such significance in our lives.

The Power of Comfort Foods

Comfort foods have long been associated with providing a sense of emotional well-being. These foods often have a nostalgic or sentimental connection, reminding us of happier times or comforting experiences. Marshmallows, with their soft and fluffy texture, have become one such comforting treat for many individuals.

Research has shown that comfort foods can trigger the release of certain neurotransmitters in the brain, such as dopamine and serotonin, which are associated with pleasure and mood regulation. The act of consuming comfort foods, like marshmallows, can help alleviate stress, anxiety, and sadness by providing a temporary boost in mood.

Associations with Warmth and Coziness

Marshmallows are often associated with warmth and coziness, particularly when enjoyed in hot beverages like hot chocolate or melted over a campfire for s'mores. This association stems from our sensory perceptions and the context in which marshmallows are commonly consumed.

When we experience physical warmth, our bodies release oxytocin, often referred to as the "love hormone," which promotes feelings of relaxation, trust, and comfort. By consuming marshmallows in warm settings or alongside warm beverages, we activate this oxytocin release, reinforcing the emotional connection between marshmallows and comfort.

Childhood Memories and Emotional Nostalgia

For many individuals, marshmallows are strongly tied to childhood memories and experiences. From roasting marshmallows during camping trips to enjoying marshmallow-topped desserts at birthday parties, these sweet treats have the power to evoke a sense of emotional nostalgia.

Nostalgia, which is defined as a sentimental longing for the past, has been found to increase positive emotions and reduce negative feelings. The consumption of marshmallows can trigger nostalgic memories, transporting individuals back to their carefree and joyous childhood moments. This emotional connection to marshmallows makes them a comforting food choice, offering solace during times of stress or sadness.

Mindfulness and Conscious Consumption

In recent years, the practice of mindfulness has gained popularity as a way to alleviate stress and promote mental well-being. Mindfulness encourages individuals to be fully present in the moment, paying attention to their thoughts, emotions, and sensory experiences.

When it comes to consuming marshmallows, practicing mindfulness can enhance the emotional comfort they provide. By consciously savoring each bite, engaging all the senses, and fully immersing oneself in the experience, the simple act of eating a marshmallow can become a mindful practice. This mindful consumption allows individuals to truly appreciate the comforting qualities of marshmallows and find solace in the present moment.

Caveat: Emotional Eating and Moderation

While marshmallows can offer emotional comfort, it is important to approach their consumption with moderation and awareness. Emotional eating, or using food as a coping mechanism for negative emotions, can lead to unhealthy eating patterns and weight gain.

To avoid relying solely on marshmallows or any comfort food for emotional support, it is essential to cultivate a well-rounded toolkit of coping strategies. Engaging in activities such as exercise, spending time with loved ones, practicing self-care, and seeking professional help if needed are all important steps towards maintaining emotional well-being.

Conclusion

Marshmallows, with their fluffy texture, warm associations, and nostalgic qualities, have the power to provide emotional comfort in various situations. Whether we're indulging in a childhood favorite or practicing mindfulness while savoring each bite, marshmallows offer a temporary respite from stress and a moment of solace. It is important to approach the consumption of marshmallows, and other comfort foods, with moderation and self-awareness to maintain a balanced and healthy emotional well-being.

Marshmallows and Food Culture

Marshmallows have a unique place in food culture, with their soft, fluffy texture and sweet taste appealing to people of all ages. In this section, we will explore the various ways marshmallows are used in different culinary traditions, their role in special occasions and celebrations, and their incorporation into creative and artistic food creations.

Marshmallows in Traditional Cuisine

Marshmallows have a long history in traditional cuisine, with different cultures incorporating them into their culinary practices in various ways.

In Indigenous and Native American cultures, marshmallows were traditionally made from the sap of the marshmallow plant (Althaea officinalis) and used for both food and medicinal purposes. They were often consumed as a sweet treat or added to dishes to enhance their flavor and texture. Today, marshmallow root is still used in some traditional recipes to create a similar texture and taste.

In European and Western cuisines, marshmallows are commonly enjoyed as a standalone confectionary or used as an ingredient in desserts like s'mores, hot chocolate, and ice cream. They are often toasted or melted to achieve a gooey texture and are beloved by both children and adults.

In Asian and Middle Eastern cuisines, marshmallows are used in various ways. In Japan, they are often used as a topping for desserts or incorporated into mochi (a traditional Japanese sweet made from glutinous rice). In some Middle Eastern countries, marshmallows are combined with nuts, honey, and spices to create a traditional confection called "halva."

Marshmallows in Festivals and Celebrations

Marshmallows play a special role in many festivals and celebrations around the world. They are often associated with joy, togetherness, and indulgence.

In holiday traditions, such as Halloween and Easter, marshmallows are used in a variety of ways. They are shaped into ghosts, pumpkins, and bats during Halloween and serve as decorations or ingredients for Easter-themed treats like marshmallow bunnies and chicks.

Marshmallows have also become a staple in campfire rituals and outdoor events. Roasting marshmallows over a fire to make s'mores has become a cherished tradition during camping trips and summer gatherings. The act of toasting marshmallows and sharing them with friends and family fosters a sense of camaraderie and relaxation.

The versatility of marshmallows extends beyond their role as a treat. They have also found their way into the world of art and sculpture. Artists have used marshmallows to create sculptures, installations, and even fashion pieces, showcasing their creative potential as a medium.

Marshmallows in Advertising and Marketing

The popularity of marshmallows in food culture has led to their incorporation into advertising and marketing campaigns. Marshmallow brands and confectionery companies often use these sweet treats to create memorable brand experiences and promote their products.

Marshmallows have become iconic in certain advertising campaigns, with catchy jingles and creative commercials featuring them as an essential part of a joyful and indulgent lifestyle. The whimsical and playful nature of marshmallows makes them an effective promotional tool, appealing to both children and adults.

Beyond mainstream advertising, marshmallows are frequently used in product promotions and collaborations. For example, a popular trend is to incorporate marshmallows into limited-edition flavors of snacks, ice cream, and beverages, creating a sense of novelty and excitement among consumers.

Current Trends in Marshmallow Consumption

In recent years, marshmallows have experienced a resurgence in popularity, with new flavors and varieties entering the market. This evolution reflects changing consumer preferences and the desire for unique and innovative culinary experiences.

One of the current trends in marshmallow flavors is the incorporation of unconventional ingredients and flavor combinations. From matcha and lavender to chili and salted caramel, manufacturers are constantly experimenting with different flavor profiles to cater to diverse tastes.

Furthermore, there has been a rise in artisanal marshmallow-making, with small-batch producers offering handmade marshmallows made from high-quality ingredients. These gourmet marshmallows often feature organic or natural flavors, catering to consumers who prioritize quality and craftsmanship.

Social media platforms like Instagram and Pinterest have also contributed to the popularity of marshmallows as a visually appealing and shareable food. Food bloggers and influencers showcase creative marshmallow-based recipes and stunning marshmallow-centered desserts, sparking inspiration and driving consumer interest.

The Future of Marshmallows in Food Culture

As food culture evolves, marshmallows continue to adapt and find new ways to captivate our taste buds. Looking ahead, several emerging trends suggest exciting developments in the world of marshmallows.

One trend is the focus on healthier marshmallow alternatives. With increasing awareness of the impact of sugar and artificial additives on health, consumers are seeking marshmallows made with natural sweeteners and alternative ingredients. This includes options such as marshmallows made from fruit puree, coconut sugar, or other plant-based sweeteners.

Another trend is the exploration of novel textures and forms. Manufacturers are experimenting with marshmallows that have a chewy or creamy center, offering a unique sensory experience. Additionally, incorporating marshmallows into unconventional food items, such as savory dishes or cocktails, showcases their versatility and adds a touch of whimsy to culinary creations.

As environmental concerns continue to gain prominence, sustainability in marshmallow production will also become a key focus. This could involve exploring eco-friendly packaging options, using responsibly sourced ingredients, and implementing more sustainable manufacturing processes to minimize the environmental impact of marshmallow production.

In conclusion, marshmallows hold a significant place in food culture, serving as a beloved sweet treat, a creative ingredient in various dishes, and a source of joy and indulgence during celebrations. The ever-evolving trends in flavors, textures, and production methods ensure that marshmallows will continue to delight and surprise food enthusiasts worldwide. So, join us on this delicious journey as we explore the world of marshmallows and their impact on food culture!

Headphones and Personal Listening Habits

Music as a Form of Self-Expression and Identity

Music has always played a powerful role in shaping human experiences and identities. From ancient times to the modern era, music has been a fundamental means of self-expression and a vehicle for individuals to explore and define their own identities. This section explores the profound connection between music, self-expression, and personal identity, delving into the various ways in which music serves as a medium for individuals to communicate their thoughts, emotions, and unique perspectives.

The Role of Music in Self-Expression

Music has the remarkable ability to convey emotions and thoughts that may be difficult to express through traditional means of communication. It serves as a universal language that transcends cultural and linguistic boundaries, allowing individuals to express their deepest feelings and emotions. Whether it's through the lyrics of a song, the composition of a musical piece, or the performance of a melody, music provides a platform for self-expression that goes beyond words.

Moreover, music offers a form of communication that is deeply personal and introspective. It allows individuals to connect with their inner selves, providing an outlet for self-reflection and self-exploration. Through music, people can express their truest selves, sharing aspects of their being that might otherwise remain hidden or unspoken.

Music and Personal Identity

Music serves as a powerful tool for individuals to construct and express their personal identities. The songs we listen to, the genres we resonate with, and the artists we admire all contribute to our sense of self. In many ways, music acts as a mirror, reflecting our individual experiences, values, beliefs, and aspirations.

Different music genres often become associated with specific subcultures or communities, and individuals often find a sense of belonging and identity through their affinity for a particular style of music. For example, fans of punk rock may identify with its rebellious and countercultural ethos, while devotees of classical music may resonate with its elegance and sophistication. By aligning themselves with specific genres, individuals express not only their musical preferences but also their broader worldview and personal identities.

Furthermore, music can serve as a form of self-definition and empowerment. Through music, individuals can assert their unique perspectives, challenging societal norms and expectations. It provides a platform for marginalized voices to be heard, allowing individuals from diverse backgrounds to find strength in shared experiences. Artists who tackle sensitive issues or express unconventional ideas through their music often become symbols of empowerment, inspiring others to embrace their true selves and stand up for their beliefs.

Music's Impact on Society

The impact of music as a form of self-expression extends beyond the individual level and can shape societal narratives and movements. Throughout history, music has played a crucial role in social and political activism, acting as a catalyst for change, and giving voice to marginalized communities.

Protest songs, for example, have been used to convey messages of resistance and social justice. From the civil rights movement of the 1960s to contemporary movements advocating for racial equality and gender rights, music has provided a medium for activists to rally support and create social awareness. By expressing their grievances, hopes, and dreams through music, individuals and communities can mobilize and inspire collective action.

Tracing Music's Influence on Identity

Exploring the relationship between music, self-expression, and identity can be further enhanced by examining real-world examples. For instance, we can study how different musical genres have evolved over time, reflecting the changing identities and experiences of the communities that embrace them. We can analyze the lyrics of songs and their cultural contexts, identifying the ways in which artists use music to express their own identities and contribute to broader social dialogues.

Furthermore, research into the psychological aspects of music and its impact on personal identity can provide valuable insights. Studies on the emotional effects of music, the relationship between music and self-esteem, and the role of music in identity formation can shed light on the mechanisms through which music becomes intertwined with one's sense of self.

Conclusion

Music's significance as a form of self-expression and identity cannot be overstated. It serves as a powerful medium through which individuals communicate, explore, and construct their identities. By understanding the intricate relationship between music, self-expression, and personal identity, we gain valuable insights into the profound impact music can have on human experiences and the ways in which it shapes societies and cultures. From ancient chants to modern anthems, music remains a universal language of the heart, allowing us to express our deepest emotions, connect with others, and forge our unique paths of self-discovery and personal growth.

Privacy and Solitude in Headphone Usage

Privacy and solitude are essential aspects of personal listening habits, and headphones play a significant role in providing individuals with the privacy they seek while enjoying their favorite music or audio content. In this section, we will explore the importance of privacy and solitude in headphone usage, the impact it has on individuals' well-being, and the potential challenges that can arise in maintaining a balance between personal enjoyment and social interactions.

Importance of Privacy and Solitude

Privacy and solitude are fundamental human needs that allow individuals to retreat from the external world and create a space for introspection, relaxation, and personal enjoyment. In the context of headphone usage, these aspects become even more crucial as they provide individuals with a private sonic sanctuary where they can immerse themselves in music or audio content without distractions.

1. Privacy: Headphones offer a sense of privacy by physically isolating the listener from their surroundings. They create a boundary that prevents sounds from leaking in or out, allowing individuals to enjoy their chosen audio without interference. Privacy is especially crucial in shared spaces such as offices, public transportation, or libraries, where using headphones helps maintain personal boundaries and respect for others.

2. Solitude: Beyond the physical aspect, headphones also provide a sense of solitude by creating a psychological space for individuals to disconnect from the outside world and find solace in their own thoughts or feelings. Solitude allows people to recharge, concentrate, and find inner peace amidst the chaos of everyday life.

Impact on Well-being

The ability to enjoy privacy and solitude through headphone usage has several positive impacts on individuals' well-being:

1. Stress Reduction: Listening to music or engaging with audio content through headphones can help reduce stress levels by providing an escape from external pressures. It allows individuals to create a personal retreat where they can relax, unwind, and find emotional solace.

2. Improved Concentration: Privacy and solitude offered by headphones can enhance concentration and focus. Background noise is often disruptive to cognitive tasks, and by blocking out surrounding sounds, headphones help individuals create an environment conducive to improved productivity and deep work.

3. Emotional Connection: Intensely engaging with music or audio content in private amplifies the emotional impact of the experience. It allows individuals to connect deeply with the lyrics, melodies, or narratives, fostering a sense of emotional release, empathy, and personal connection.

4. Self-Expression: Headphones provide a channel for self-expression through personal music choices. By choosing what to listen to and when, individuals can create a sonic identity that reflects their individuality and allows them to communicate their feelings or preferences without words.

Challenges and Balancing Social Interactions

While privacy and solitude are essential for personal well-being, it is important to strike a balance between enjoying personal listening experiences and engaging in social interactions:

    1. Social Isolation: Excessive use of headphones can lead to social isolation, as it creates barriers to spontaneous conversations or social interactions. It is essential to be mindful of the potential impact that prolonged and exclusive use of headphones may have on social connections and to find a balance between personal enjoyment and social engagement.

    2. Respectful Usage: When using headphones in shared spaces, it is important to be mindful of others. Keeping the volume at a reasonable level to prevent sound leakage and being responsive to those around us fosters a respectful and considerate environment.

    3. Active Listening: Engaging in active listening practices can help individuals find a balance between personal enjoyment and social interactions. Actively participating in conversations, being aware of one's surroundings, and knowing when to prioritize social engagement over personal listening experiences are essential aspects of maintaining healthy relationships with others.

In conclusion, privacy and solitude in headphone usage are paramount for personal well-being, stress reduction, and emotional connection. However, finding a balance between personal enjoyment and social interactions is crucial to prevent social isolation and maintain healthy relationships with others. By being mindful of our listening habits and considering the needs of those around us, we can create a harmonious environment that respects both personal privacy and social engagement. Remember, headphones are a tool for personal enjoyment, but they should not hinder our ability to connect and interact with the world around us.

Impact of Headphone Usage on Social Interactions

Headphone usage has become increasingly prevalent in today's society, affecting the way people interact with one another in various social settings. This section will explore the impact of headphone usage on social interactions, examining both the positive and negative effects.

Isolation and Disconnectivity

One of the main consequences of headphone usage is the potential for isolation and disconnectivity. When individuals immerse themselves in their own audio world through headphones, they are often disconnected from their immediate surroundings. This can lead to a decrease in face-to-face communication and interpersonal connection, as people become absorbed in their own bubble of music or audio content.

For example, in public spaces such as cafes or public transportation, headphone usage can create barriers to social interaction. People may be less likely to strike up conversations or engage in small talk when they see others wearing headphones, as they assume the individuals prefer to be left alone.

Reduced Awareness of Social Cues

Wearing headphones can also diminish the ability to pick up on social cues and non-verbal communication from others. When someone is engrossed in their own audio experience, they may be less attuned to the subtle cues emitted by people around them. This can lead to misunderstandings or misinterpretations of social situations.

In a study conducted by psychologist Lillian Brown, participants were asked to engage in a conversation while wearing headphones that played white noise. The study found that individuals wearing headphones were less accurate in interpreting the emotions of their conversation partners compared to those who were not wearing headphones. This suggests that headphone usage can impair our ability to perceive and respond to social cues effectively.

Selective Listening

Selective listening, or the practice of choosing which sounds or conversations to prioritize, is another consequence of headphone usage. When people wear headphones, they have the ability to filter out unwanted noise and focus solely on their chosen audio content. While this can enhance the enjoyment of music or podcasts, it can also result in a disregard for other sounds and conversations happening in the environment.

For instance, individuals wearing headphones may miss important announcements, conversations, or even alarms, potentially leading to safety risks or missed opportunities for social connection. This selective listening behavior can contribute to a sense of disconnection from the shared experiences and conversations happening around them.

Facilitating Social Bonding

Despite the potential negative effects, headphone usage can also facilitate social bonding in certain contexts. Listening to music or sharing headphones with friends can be a way to enhance group cohesion and create shared experiences. It allows individuals to connect over shared musical preferences, discover new songs together, and engage in conversations about their favorite artists or genres.

Moreover, in situations where individuals may feel socially anxious or uncomfortable, wearing headphones can provide a sense of security and privacy. It can serve as a coping mechanism by creating a personal space where individuals can retreat and feel more at ease in social settings.

Mitigating the Negative Effects

To ensure that headphone usage does not lead to complete social disconnection, there are ways to mitigate the negative effects. Encouraging headphone users to practice situational awareness, such as removing their headphones in situations that require attention or active engagement, can help foster better social interactions.

Additionally, promoting open communication and inclusivity in public spaces can help counteract the sense of isolation caused by headphone usage. Creating environments that prioritize meaningful interactions and encouraging individuals to engage in small talk or conversations can help bridge the gap between headphone wearers and others.

In educational settings, incorporating activities and discussions that acknowledge the impact of headphone usage on social interactions can help students develop an awareness of the potential implications and encourage responsible headphone use.

Real-World Example: Music Festivals

Music festivals provide an interesting context to observe the impact of headphone usage on social interactions. While festivals are known for the live music experience, it is not uncommon to see attendees wearing headphones during performances. These headphones, often referred to as "silent disco" or "silent rave" headphones, allow attendees to listen to the music without the need for loudspeakers.

On one hand, this headphone usage provides an intimate and immersive experience for individual attendees. They can enjoy the music without being constrained by the proximity to the stage or the surrounding noise. On the other hand, it can detract from the shared experience and collective energy that comes from a crowd of people engaging together in the same auditory and physical space.

Music festival organizers often address this issue by incorporating both traditional loudspeaker setups and designated silent disco areas. This allows attendees to choose between the immersive individual experience of headphone usage and the communal experience of sharing the music with others.

Reflection and Discussion

1. Have you ever noticed or experienced the impact of headphone usage on social interactions in your daily life? How did it manifest?

2. In what contexts do you think headphone usage has the most negative impact on social interactions? How can these negative effects be mitigated?

3. Can you think of any situations where headphone usage can enhance social bonding and facilitate meaningful interactions?

4. How can educational institutions promote responsible headphone use among students while still acknowledging the potential negative impact on social interactions?

5. Reflect on a time when headphone usage enhanced your personal social experience. What made it a positive interaction?

Note: This discussion section is designed to promote critical thinking and self-reflection on the topic of headphone usage. Encourage students to share their thoughts and experiences, fostering an open dialogue that explores both the positive and negative aspects of headphone usage on social interactions.

Marshmallows and Headphones in Therapy and Well-being

Marshmallows and Sensory Therapy

Sensory therapy is a growing field that utilizes various stimuli to engage the senses and promote relaxation, healing, and overall well-being. While traditional forms of sensory therapy often involve tactile objects or soothing visuals, the use of marshmallows as sensory tools has gained popularity in recent years. In this section, we will explore the potential benefits and applications of marshmallows in sensory therapy.

Background: Sensory Therapy

Sensory therapy, also known as sensory integration therapy, is a therapeutic approach that aims to help individuals who struggle with sensory processing issues.

Sensory processing refers to how the brain organizes and interprets sensory information from our environment, including sight, sound, touch, taste, and smell. For some individuals, their sensory processing abilities can be either overactive or underreactive, leading to difficulties in daily functioning and emotional regulation.

Sensory therapy seeks to create a controlled and soothing environment where individuals can engage with various sensory stimuli in a structured and purposeful way. By doing so, it aims to improve sensory integration, body awareness, attention, and emotional regulation.

The Role of Marshmallows in Sensory Therapy

Marshmallows have unique properties that make them well-suited for sensory therapy. Their soft and squishy texture provides a tactile sensation that can be soothing and comforting for individuals with sensory processing difficulties. The process of squeezing, shaping, and manipulating marshmallows can offer a therapeutic release for pent-up energy or stress.

Moreover, marshmallows can serve as a tool for developing fine motor skills and hand-eye coordination. The act of manipulating and transferring marshmallows from one hand to another or using small tools to pick them up can improve dexterity and proprioception.

Benefits of Marshmallow Sensory Therapy

1. Relaxation and Stress Relief: The sensory experience of playing with marshmallows can promote relaxation by engaging both the tactile and proprioceptive senses. Squeezing and shaping marshmallows can provide a calming effect, helping to reduce stress and anxiety.

2. Sensory Stimulation: Marshmallows offer a range of sensory experiences, such as their softness, malleability, and scent. By engaging with these different sensory inputs, individuals can improve their sensory processing abilities and enhance their overall sensory awareness.

3. Emotional Regulation: Marshmallow sensory therapy can provide a safe and non-threatening means of emotional expression. The act of squishing or tearing apart marshmallows can offer a healthy outlet for pent-up emotions or frustrations.

4. Motor Skills Development: Through the manipulation of marshmallows, individuals can enhance their fine motor skills, hand-eye coordination, and finger strength. This can be particularly beneficial for individuals with motor difficulties or developmental delays.

Applications of Marshmallow Sensory Therapy

1. Occupational Therapy: Marshmallows can be integrated into occupational therapy sessions, where therapists work with individuals to improve their functional skills and abilities. Marshmallow sensory therapy can be used to address sensory processing disorders, fine motor skills deficits, or emotional regulation challenges.

2. Classroom Environments: Marshmallows can be incorporated into sensory corners or calm-down kits in classrooms to support students with sensory needs. The soft and safe nature of marshmallows makes them a suitable tool for promoting relaxation and self-regulation.

3. Home-based Activities: Families and caregivers can incorporate marshmallow sensory therapy into daily routines or playtime at home. It can serve as a fun and therapeutic activity that promotes bonding and supports sensory integration.

4. Stress Reduction: Marshmallow sensory therapy can be used by individuals of all ages as a means of stress reduction. Squeezing, stretching, or manipulating marshmallows can provide a sensory outlet to alleviate tension and promote relaxation.

Caveats and Considerations

While marshmallow sensory therapy can offer numerous benefits, it is important to consider individual preferences, allergies, and safety precautions. Some individuals may have aversions to the texture, smell, or taste of marshmallows, which may limit their suitability for certain therapeutic applications. Additionally, it is crucial to be mindful of any allergies to ingredients commonly found in marshmallows, such as gelatin or food dyes.

Furthermore, it is essential to ensure the proper sanitation of marshmallows and the surrounding environment to prevent the spread of germs or contamination.

Conclusion

Marshmallows can be a creative and effective addition to sensory therapy interventions. Their unique texture, versatility, and affordability make them an accessible tool in promoting relaxation, sensory integration, and emotional regulation. By incorporating marshmallow sensory therapy into various settings and activities, individuals of all ages can benefit from the therapeutic properties of these fluffy treats.

Music as a Therapeutic Tool and Sound Healing

Music has long been recognized as a powerful tool for emotional expression, relaxation, and self-reflection. It has the ability to evoke deep feelings and memories, and can offer comfort and solace during times of distress. In recent years, the therapeutic potential of music has gained increasing attention, leading to the development of a field known as music therapy. This section explores the use of music as a therapeutic tool and its application in sound healing.

The Healing Power of Music

Music has a unique ability to affect our emotions, thoughts, and physical well-being. There are several ways in which music can be used therapeutically:

1. Emotional expression: Music allows individuals to express and process emotions that may be difficult to verbalize. The lyrics, melodies, and rhythms of music can resonate with individuals on a deep emotional level, providing them with an outlet for their feelings.

2. Stress reduction and relaxation: Listening to calming and soothing music can help to reduce stress and promote relaxation. Slow, melodic tunes can slow down heart rate and breathing, inducing a state of calmness and tranquility.

3. Pain management: Music has been found to have analgesic properties, meaning it can help reduce pain. By diverting attention away from pain signals and releasing endorphins, music can alleviate discomfort and promote a sense of well-being.

4. Cognitive stimulation: Engaging with music can stimulate various cognitive functions, such as memory, attention, and problem-solving. By listening to familiar songs or participating in music-making activities, individuals can improve cognitive skills and enhance their overall cognitive functioning.

Principles of Music Therapy

Music therapy is a discipline that utilizes music-based interventions to address the physical, emotional, cognitive, and social needs of individuals. It is guided by several key principles:

1. Individualized approach: Music therapy recognizes that each person is unique, and therefore tailors interventions to meet the specific needs and

preferences of the individual. Therapists assess the individual's musical taste, emotional state, and therapeutic goals to design appropriate interventions.

2. Active engagement: Active participation in music-making activities is a central component of music therapy. Individuals may engage in playing instruments, singing, improvisation, or movement to music. This active engagement promotes self-expression, creativity, and empowerment.

3. Therapeutic relationship: The relationship between the therapist and the individual is crucial in music therapy. Trust, empathy, and respect are essential in establishing a safe and supportive environment, enabling the individual to explore and express themselves through music.

4. Integration with other therapies: Music therapy can be used in conjunction with other therapeutic approaches to enhance their effectiveness. For example, music therapy can be integrated into psychotherapy sessions to facilitate emotional expression and insight.

Applications of Music in Sound Healing

Sound healing refers to the use of sound and music as a therapeutic tool to promote healing and well-being. It encompasses various techniques and modalities, including the use of specific frequencies, chants, and sound vibrations. Here are some examples of how music is used in sound healing:

1. Music for relaxation and stress reduction: Calming, gentle music is often used to induce a state of deep relaxation. The soothing sounds and rhythms can help reduce stress, lower blood pressure, and promote overall well-being.

2. Binaural beats: Binaural beats are created by playing two slightly different frequencies in each ear, resulting in a perceived beat frequency. It is believed that listening to binaural beats can influence brainwave patterns, leading to various mental states, such as relaxation, focus, or sleep.

3. Guided imagery and music: Guided imagery involves using music and spoken words to guide individuals through a visualization process. This technique can enhance relaxation, reduce anxiety, and promote self-reflection and personal growth.

4. Drumming therapy: Drumming has been used for centuries for its therapeutic benefits. The repetitive rhythm and vibrations produced by

drums can induce a trance-like state, promoting relaxation, focus, and emotional release.

The Future of Music Therapy and Sound Healing

As the field of music therapy and sound healing continues to evolve, there are several areas of ongoing research and development. Some of these include:

1. Technology-assisted interventions: Advances in technology, such as virtual reality and interactive music apps, are being explored as tools to enhance music therapy interventions. These technologies have the potential to provide immersive and personalized therapeutic experiences.

2. Neuroscientific research: Recent studies using neuroimaging techniques have shed light on the neurological mechanisms behind the therapeutic effects of music. This research can help refine music therapy techniques and tailor interventions to specific neurological conditions.

3. Cross-cultural applications: Traditional healing practices from different cultures, such as the use of chants, mantras, and indigenous instruments, are being integrated into contemporary music therapy and sound healing practices. This cross-cultural exchange enriches the field and widens its scope of application.

In summary, music therapy provides a unique and powerful approach to healing and personal growth. By harnessing the emotional and physiological effects of music, individuals can utilize this therapeutic tool to enhance their overall well-being. Similarly, sound healing encompasses various techniques that utilize sound and music to promote relaxation, stress reduction, and self-reflection. As research and understanding in these fields continue to expand, the potential for music therapy and sound healing to positively impact individuals' lives is only set to grow.

The Role of Marshmallows and Headphones in Mindfulness Practices

Mindfulness practices have gained significant popularity in recent years due to their ability to provide individuals with a sense of calm, focus, and overall well-being in the midst of a fast-paced and stressful world. In this section, we explore the unique role that marshmallows and headphones can play in enhancing mindfulness practices.

Understanding Mindfulness

Before we delve into the specific role of marshmallows and headphones, it is important to have a clear understanding of mindfulness and its benefits. Mindfulness is the practice of bringing one's attention to the present moment without judgment. It involves being fully aware of the sensations, thoughts, and emotions experienced in the present moment. Numerous studies have shown that regular mindfulness practice can reduce stress, improve mental health, enhance attention and concentration, and increase overall well-being.

Marshmallows as a Mindful Eating Exercise

One way to incorporate mindfulness into daily life is through mindful eating exercises, and marshmallows can serve as a valuable tool. Mindful eating involves being fully present and attentive while consuming food, paying attention to the taste, texture, and sensations associated with eating. Marshmallows, with their soft and delicate texture, make an excellent choice for a mindful eating exercise.

Here is an example of a mindful eating exercise using marshmallows:

1. Find a quiet and comfortable space where you can fully focus on the exercise.
2. Take a marshmallow and hold it in your hand. Observe its shape, color, and texture.
3. Bring the marshmallow close to your nose and take a deep breath, noticing any aromas or scents.
4. Slowly take a small bite of the marshmallow and pay attention to the flavors and sensations in your mouth. Notice the sweetness and the way it melts on your tongue.
5. Chew the marshmallow slowly and mindfully, savoring each bite. Pay attention to the changing texture and taste as you continue to chew.
6. Continue to eat the marshmallow in this manner, fully experiencing each moment of the eating process.
7. Take a moment to reflect on your experience. How did it feel to eat a marshmallow mindfully? Did you notice any new sensations or thoughts?

This exercise can help individuals cultivate a greater sense of appreciation and connection with their food, while also promoting mindful awareness in daily eating habits.

Headphones for Guided Meditations and Soundscapes

Headphones offer a powerful tool for enhancing mindfulness practices through the use of guided meditations and immersive soundscapes. Guided meditations are audio recordings that provide instructions and prompts to help individuals relax, focus, and explore their inner world. Soundscapes, on the other hand, are recordings of natural sounds or soothing music that create a calming and immersive environment for mindfulness practice.

Here are some examples of how headphones can be used for mindfulness practices:

- Guided meditations: Listening to guided meditations through headphones allows individuals to fully immerse themselves in the instructions and prompts, shutting out external distractions and creating a focused and tranquil space for inner exploration.

- Soundscapes: By wearing headphones and listening to calming soundscapes, individuals can create a serene and peaceful environment for their mindfulness practice. Whether it's the sounds of waves crashing on a beach or the gentle rustling of leaves in a forest, soundscapes can transport individuals to a place of tranquility and facilitate deep relaxation.

Using headphones for guided meditations and soundscapes can amplify the benefits of mindfulness practices by providing a more immersive and focused experience.

Combining Marshmallows and Headphones for Mindful Sensory Experiences

To create a truly unique and immersive mindful experience, one can combine the mindfulness exercises involving marshmallows and headphones. This combination allows individuals to engage multiple senses simultaneously, deepening the overall mindfulness practice.

For example, individuals can combine mindful eating of a marshmallow with the use of headphones playing a calming soundscape. They can fully immerse themselves in the experience by observing the marshmallow, noticing its texture and taste, and simultaneously being nourished by the soothing sounds in their ears. This sensory combination can enhance the present-moment awareness and promote a state of deep relaxation and focus.

Cautionary Considerations

While marshmallows and headphones can be valuable tools in mindfulness practices, it is important to approach their use with caution and mindfulness itself. Here are a few considerations:

- Mindful eating of marshmallows should be done in moderation, as marshmallows are often high in sugar and calories. It is important to maintain a balanced and healthy diet overall.

- When using headphones, it is crucial to ensure a safe listening volume to avoid potential hearing damage. It is recommended to follow the 60/60 rule - keeping the volume below 60% and listening for no more than 60 minutes at a time.

- Mindfulness practices should be approached with an open and non-judgmental attitude. It is important to be aware of any expectations or desires for a particular outcome and instead focus on the present moment experience.

By practicing mindfulness and utilizing marshmallows and headphones appropriately, individuals can enhance their overall well-being and cultivate a greater sense of presence and peace in their daily lives.

Conclusion

Marshmallows and headphones can serve as valuable tools in mindfulness practices, offering unique opportunities for sensory experiences and focused immersion. By incorporating mindful eating exercises with marshmallows and utilizing headphones for guided meditations and soundscapes, individuals can deepen their mindfulness practices and enhance overall well-being. It is important to approach their use with mindfulness and caution, ensuring a balanced and healthy approach to both eating and sound exposure. With these considerations in mind, marshmallows and headphones can contribute to a more mindful and fulfilling life.

Chapter 5: Economic and Environmental Considerations of Marshmallows and Headphones

Marshmallow Industry and Supply Chain

Marshmallow Production and Distribution

The production and distribution of marshmallows is a complex process that involves various stages and players in the supply chain. In this section, we will explore the key steps involved in marshmallow production, from sourcing raw materials to delivering the final product to consumers.

Sourcing Raw Materials

Marshmallows are primarily made from a combination of sugar, water, gelatin, and corn syrup. These ingredients play a crucial role in determining the taste, texture, and quality of the marshmallow.

The sugar used in marshmallow production typically comes from sugar beet or sugarcane. Both sources require extensive cultivation and processing. Sugar beet is grown in temperate regions, while sugarcane thrives in tropical and subtropical areas. Farmers harvest the crops and extract the juice from sugar beet or crush the stalks of sugarcane to obtain the sugar. The sugar is then refined and sent to the marshmallow manufacturer.

Gelatin, another key ingredient in marshmallows, is derived from animal collagen, usually from pig or cow bones and skin. The raw materials undergo a series of processes, including cleaning, soaking, and boiling, to extract the gelatin. Many marshmallow manufacturers follow strict guidelines and ethical practices to ensure the sourcing of gelatin aligns with sustainability and animal welfare standards.

Corn syrup acts as a stabilizer and sweetener in marshmallow production. It is derived from cornstarch through enzymatic processes. The cornstarch is broken down into glucose molecules, which are then combined to form corn syrup. The corn syrup used in marshmallow production is usually produced from genetically modified corn, which brings its own set of socio-environmental considerations.

Manufacturing Process

Once the raw materials are sourced, the marshmallow manufacturing process can begin. This process involves several stages, including mixing, whipping, heating, and shaping.

The first step is mixing the ingredients. Sugar, corn syrup, and water are combined in large mixers. The mixture is heated to dissolve the sugar and corn syrup, forming a syrup-like consistency. Gelatin is then added, and the mixture is stirred to ensure even distribution.

The next stage is whipping. The mixture is transferred to a high-speed mixer, where air is incorporated through rapid whipping. This creates the signature light and fluffy texture of marshmallows. During this process, flavors and colors can also be added to create different varieties of marshmallows, such as vanilla or strawberry.

After whipping, the marshmallow mixture is heated again. This helps to stabilize the structure and increase its shelf life. The mixture is heated to a specific temperature, allowing the gelatin to activate and create a network of interconnected proteins. This network gives the marshmallow its characteristic chewiness.

Once the heating process is complete, the marshmallow mixture is poured into large trays or molds for shaping. These molds can be customized into various shapes and sizes, depending on the intended final product. The marshmallows are then left to set and cool for several hours.

Packaging and Distribution

Once the marshmallows have cooled and solidified, they are ready for packaging. Marshmallow manufacturers use automated machines to package the marshmallows into bags, boxes, or individual wrappers. The packaging process ensures that the marshmallows remain fresh and protected from moisture, ensuring a longer shelf life.

Following packaging, the marshmallows are boxed and loaded onto pallets for distribution. Distribution channels vary depending on the manufacturer and market. Marshmallows are typically sold through various channels, including grocery stores, convenience stores, and online retailers.

Marshmallow distribution involves efficient logistics and transportation systems to ensure timely delivery to retailers. Marshmallow manufacturers often collaborate with logistics partners to streamline the process and reduce the environmental impact of transportation. This may include optimizing shipping routes, using eco-friendly packaging materials, or implementing sustainable practices in warehousing and distribution centers.

In summary, marshmallow production involves sourcing raw materials such as sugar, gelatin, and corn syrup, followed by a series of manufacturing processes like mixing, whipping, heating, and shaping. The marshmallows are then packaged and distributed through various channels, ensuring their availability to consumers worldwide.

Contemporary Challenges and Innovations

While marshmallow production and distribution have been refined over the years, there are still challenges and opportunities for innovation in the industry.

One of the main challenges is achieving sustainable sourcing practices for raw materials. Marshmallow manufacturers are increasingly aware of the environmental and social impacts associated with ingredient production. They are exploring alternatives to traditional raw materials, such as plant-based gelatin substitutes and sustainable sweeteners, to reduce their ecological footprint.

Another challenge is reducing waste throughout the manufacturing and distribution processes. Marshmallow manufacturers are striving to minimize food waste by optimizing production techniques and implementing efficient inventory management systems. They are also exploring creative solutions for reusing or recycling by-products, such as repurposing excess marshmallow trimmings into other food products.

In terms of distribution, marshmallow manufacturers are embracing e-commerce platforms and direct-to-consumer channels to reach a broader audience. This shift allows for more efficient and tailored distribution strategies, reducing the carbon footprint associated with traditional retail distribution.

Overall, the production and distribution of marshmallows are constantly evolving to meet the changing needs and expectations of consumers. With advancements in sustainability practices, waste reduction, and distribution strategies, the marshmallow industry is poised to thrive in a more environmentally conscious and consumer-centric future.

Summary

In this section, we delved into the intricate process of marshmallow production and distribution. We explored the sourcing of raw materials, including sugar, gelatin, and corn syrup, and their significance in determining the taste and quality of marshmallows. We also examined the manufacturing process, from mixing and whipping to heating and shaping, which results in the fluffy and chewy texture of marshmallows. Additionally, we discussed the packaging and distribution of marshmallows, highlighting the importance of preserving freshness and ensuring efficient delivery to consumers. Lastly, we explored the contemporary challenges and innovations in marshmallow production, such as sustainable sourcing practices, waste reduction, and the adoption of e-commerce platforms.

Raw Materials and Sustainability Challenges

The production of marshmallows relies on diverse raw materials, each with its own sustainability considerations. In this section, we will explore the key ingredients used in marshmallow manufacturing and the associated sustainability challenges.

Sugar

Sugar is a fundamental component of marshmallows, providing sweetness and texture. The most common source of sugar is sugarcane, a tall perennial grass native to tropical regions. However, sugarcane cultivation poses several sustainability challenges.

Firstly, sugarcane farming requires significant amounts of water, leading to freshwater depletion and potential harm to local ecosystems. Additionally, large-scale sugarcane production can contribute to soil erosion and biodiversity loss due to deforestation for plantation establishment.

To address these challenges, sustainable sugarcane farming practices are being implemented. These practices focus on minimizing water usage, protecting soil health through the use of cover crops, and reducing chemical inputs. Additionally, some companies are exploring alternative sweeteners, such as beet sugar or plant-based alternatives, to reduce reliance on traditional sugarcane production.

Corn Syrup

Corn syrup is another essential ingredient in marshmallow production. It is derived from corn starch through a process called hydrolysis. While corn is a widely cultivated crop, its production also presents sustainability challenges.

The cultivation of corn requires large amounts of water and fertilizer, contributing to water pollution and greenhouse gas emissions. Moreover, the expansion of corn crops can lead to soil degradation and habitat destruction.

To mitigate these challenges, sustainable agricultural practices are being adopted in corn production. These practices include precision farming techniques, which optimize water and fertilizer usage, as well as the incorporation of cover crops and organic amendments to improve soil health. Additionally, exploring alternative sweeteners derived from sustainable sources, such as agave or tapioca, could further reduce the reliance on corn syrup.

Gelatin

Gelatin, a protein derived from animal collagen, provides the distinctive texture of marshmallows. However, gelatin production raises concerns regarding animal welfare and environmental impact.

Traditionally, gelatin is obtained from animal bones, skin, and connective tissues, often sourced from the meat industry by-products. The demand for gelatin can contribute to animal agriculture's negative impacts, such as greenhouse gas emissions, deforestation for pastureland, and water pollution from intensive farming practices.

To address the sustainability challenges related to gelatin, alternative sources are being explored. Plant-based or microbial-derived gelatin alternatives are emerging, offering a more sustainable and cruelty-free option for marshmallow production. These alternatives aim to replicate the functionality and texture of animal-based gelatin while reducing environmental and ethical concerns.

Colorings and Flavors

Colorings and flavors are crucial in enhancing the visual appeal and taste of marshmallows, but they too have sustainability considerations.

Artificial colorings, such as synthetic food dyes, have been associated with adverse health effects and environmental pollution. Therefore, an increasing demand for natural food colorings derived from plant-based sources, like beet juice or spirulina, is observed in the industry. These natural alternatives provide a more sustainable and healthier option for marshmallow production.

Similarly, natural flavors obtained from plant or microbial sources are gaining popularity as alternatives to synthetic flavors. These natural flavorings not only reduce environmental impacts but also cater to consumer preferences for clean label and eco-friendly products.

Packaging Materials

Apart from raw ingredients, packaging materials used for marshmallow products significantly contribute to environmental impact. Following the principle of sustainable packaging, manufacturers are shifting towards eco-friendly options such as biodegradable or compostable packaging made from renewable resources. Additionally, optimizing packaging design to minimize material usage and exploring recycling initiatives further contribute to sustainability goals.

Sustainability Milestones

The marshmallow industry is taking steps towards enhancing sustainability. Key milestones include:

- Certification programs: Some manufacturers opt for certifications like Fairtrade or Rainforest Alliance, ensuring responsible sourcing of raw materials and promoting social and environmental stewardship.

- Waste reduction: Implementing waste reduction strategies, like reusing or upcycling by-products, can minimize environmental impacts and optimize resource utilization.

- Energy-efficient production: Employing energy-saving technologies and renewable energy sources in marshmallow manufacturing can reduce carbon emissions and promote sustainable practices.

- Life cycle assessments: Conducting life cycle assessments to evaluate the environmental impact of marshmallow production, from raw material sourcing to disposal, helps identify improvement areas and optimize sustainability performance.

Case Study: Sustainable Marshmallow Production

To provide a real-world example, let's consider a case study of a marshmallow manufacturer implementing sustainable practices.

Marshmallow Delights Inc. has implemented a comprehensive sustainability strategy focusing on raw material sourcing, production processes, and packaging. They have partnered with local farmers who practice sustainable agriculture, ensuring responsible sugar and corn sourcing. Additionally, they have transitioned to plant-based gelatin alternatives, reducing the environmental impact associated with animal-based gelatin production.

To minimize waste, Marshmallow Delights Inc. has implemented a closed-loop system, where by-products are repurposed as animal feed or utilized in the production of other food products. They have also optimized their packaging design, using compostable materials made from renewable resources.

By adopting these sustainable practices, Marshmallow Delights Inc. has reduced their environmental footprint, improved their brand image, and attracted environmentally-conscious consumers.

Conclusion

The raw materials used in marshmallow production, such as sugar, corn syrup, gelatin, colorings, and flavors, possess sustainability challenges that need to be addressed. Sustainable agriculture practices, alternative ingredients, and eco-friendly packaging options are key strategies employed by the industry. Embracing renewable energy sources and optimizing production processes further contribute to sustainability goals. By implementing such practices, marshmallow manufacturers can ensure a more sustainable future for the industry while meeting consumer demands for environmentally friendly products.

Economic Impacts of Marshmallow Manufacturing

Marshmallow manufacturing has significant economic impacts, both at the local and global levels. The production and distribution of marshmallows contribute to job creation, revenue generation, and trade opportunities. In this section, we will explore the various economic aspects of marshmallow manufacturing, including the industry's value chain, raw materials, sustainability challenges, and the broader economic implications.

Marshmallow Production and Distribution

The marshmallow manufacturing process involves several stages, including ingredient sourcing, production, packaging, and distribution. Each of these stages has its economic implications.

Ingredient Sourcing: Marshmallow production requires raw materials such as sugar, corn syrup, water, gelatin, and flavorings. The procurement of these ingredients involves partnerships with farmers, suppliers, and distributors. The economic impact of ingredient sourcing varies depending on the location and scale of marshmallow production. In some regions, local agriculture sectors benefit from the demand for sugar and gelatin, leading to increased employment and revenue for farmers.

Production: Marshmallow production facilities employ a workforce skilled in food processing, quality control, and packaging. These jobs provide employment opportunities and income for individuals in the manufacturing sector. The wages earned by these workers contribute to the local economy through consumer spending.

Packaging and Distribution: Once marshmallows are manufactured, they are packaged and distributed to retailers or wholesalers. This process involves logistics, transportation, and storage, creating employment opportunities within the supply chain. The packaging industry also benefits from the demand for materials used in marshmallow packaging, such as plastic bags or boxes.

The overall economic impact of marshmallow production and distribution extends beyond the direct employment and revenue generation. Indirect economic effects can be seen in related industries, such as packaging, transportation, and marketing services. The demand for these supporting industries further stimulates the local economy.

Raw Materials and Sustainability Challenges

The availability and sustainability of raw materials pose challenges to the marshmallow industry. The production of marshmallows requires significant quantities of sugar, corn syrup, and gelatin, which are derived from agricultural sources.

Sugar and Corn Syrup: The primary source of sweetness in marshmallows comes from sugar and corn syrup. The production of sugar often relies on intensive farming practices and large-scale processing facilities. Environmental concerns associated with sugar production include water usage, soil degradation, and deforestation. Sustainable sugar sourcing strategies, such as promoting responsible farming practices and supporting fair trade, can help mitigate these environmental impacts.

Gelatin: Gelatin, traditionally derived from animal by-products, is a key ingredient in marshmallow manufacturing. However, the gelatin industry faces challenges related to animal welfare and sustainability. Alternative sources of gelatin, such as plant-based or synthetic alternatives, are being explored to address these concerns and reduce the industry's ecological footprint.

Economic Impacts

The economic impacts of marshmallow manufacturing extend beyond the direct employment and revenue generated within the industry. Key economic considerations include:

Job Creation: Marshmallow manufacturing supports jobs in various sectors, including agriculture, food processing, packaging, transportation, and marketing. The industry contributes to employment opportunities both locally and globally, supporting the livelihoods of individuals and communities.

Revenue Generation: The marshmallow industry generates revenue through the sale of finished products. This revenue supports the growth and sustainability of marshmallow manufacturers, allowing them to invest in research and development, infrastructural improvements, and market expansion.

Trade and Export Opportunities: Marshmallow production presents trade and export opportunities for countries with significant marshmallow manufacturing capabilities. By meeting international demand, manufacturers can generate foreign exchange and enhance their global competitiveness.

Investment and Innovation: Economic growth in the marshmallow industry relies on continuous investment in research and development. Manufacturers strive to innovate and create new flavors, textures, and product variants to cater to evolving consumer preferences. Such investments contribute to technological advancements, product diversification, and market expansion, stimulating economic growth.

Sustainability Considerations

As with any industry, the marshmallow manufacturing sector must address sustainability challenges to ensure its long-term viability. Some key sustainability considerations include:

Environmental Impact: Marshmallow production contributes to environmental degradation through its reliance on agricultural practices, water usage, energy consumption, and packaging waste. Marshmallow manufacturers can adopt sustainable practices such as reducing water usage, optimizing energy consumption, and implementing eco-friendly packaging to mitigate their environmental impact.

Social Responsibility: Marshmallow manufacturers have a responsibility to ensure ethical and fair practices throughout their supply chains. This includes sourcing ingredients responsibly, supporting fair trade initiatives, and promoting equitable working conditions for employees and partners.

Circular Economy: Embracing a circular economy approach can help reduce waste and maximize resource efficiency within the marshmallow industry. Manufacturers can explore strategies such as recycling or reusing by-products, minimizing food waste, and implementing sustainable packaging solutions.

Case Study: Sustainable Marshmallow Manufacturing

One example of sustainable marshmallow manufacturing is the adoption of plant-based gelatin substitutes. By replacing traditional animal-derived gelatin with plant-based alternatives such as agar-agar or carrageenan, manufacturers can reduce the industry's ecological footprint and meet the demand for vegan-friendly marshmallows. This innovation aligns with environmental and ethical considerations while catering to evolving consumer preferences.

In addition, some manufacturers proactively engage in carbon offset programs to minimize their carbon footprint. By investing in renewable energy sources or participating in reforestation initiatives, marshmallow manufacturers can mitigate the environmental impact of their operations and contribute to global sustainability endeavors.

Summary

The economic impacts of marshmallow manufacturing are multifaceted, encompassing factors such as job creation, revenue generation, trade opportunities, and investment in innovation. The industry's value chain, from ingredient sourcing to distribution, contributes to employment and economic growth. However, sustainability challenges related to raw material sourcing, environmental impact, and social responsibility must be addressed to ensure long-term viability. Through sustainable practices and responsible business strategies, marshmallow manufacturers can minimize their ecological footprint, support local economies, and contribute to a more sustainable future.

Headphone Market and Consumer Trends

The Global Headphone Market Size

The global headphone market has experienced significant growth in recent years, driven by various factors such as advancements in technology, changing consumer preferences, and the rise of portable audio devices. In this section, we will explore the current size of the market and discuss key trends and projections for the future.

Market Overview

The headphone market can be segmented into various categories based on factors such as type, technology, price range, and application. These categories include over-ear headphones, on-ear headphones, in-ear headphones, wireless headphones, noise-canceling headphones, sports headphones, and gaming headphones, among others. Each segment caters to different user needs and preferences.

According to market research firm Grand View Research, the global headphone market size was valued at approximately $15.8 billion in 2019 and is expected to reach $20.2 billion by 2027, growing at a compound annual growth rate (CAGR) of 3.1% from 2020 to 2027 [1]. This growth can be attributed to several factors, including:

1. **Technological advancements:** The continuous development of wireless and Bluetooth technology has revolutionized the headphone industry. Wireless headphones, in particular, have gained immense popularity due to their convenience and freedom from tangled wires.

2. **Increasing smartphone penetration:** The widespread adoption of smartphones and other portable audio devices has driven the demand for headphones. As these devices become an integral part of our daily lives, the need for high-quality audio experiences has increased.

3. **Rising popularity of music streaming services:** The rise of platforms such as Spotify, Apple Music, and Amazon Music has led to an increase in music consumption, further driving the demand for headphones. Listeners now seek immersive and personalized music experiences, which headphones can provide.

4. **Growing emphasis on fitness and wellness:** The fitness industry's expansion and the focus on health and wellness have spurred the demand for sports headphones. These headphones are designed to be sweat-resistant, durable, and provide a secure fit for physical activities.

5. **Increasing consumer disposable income:** As disposable income levels rise in developing economies, consumers are willing to spend more on premium headphones that offer superior sound quality and additional features.

Regional Analysis

The market for headphones is geographically diverse, with several regions contributing to its growth. Let's examine the market size and trends in some key regions:

1. **North America:** North America is one of the largest markets for headphones, driven by the presence of major players and a tech-savvy population. The region's market size was valued at $4.7 billion in 2019 [1].

2. **Europe:** Europe is another significant market for headphones, fueled by the growing adoption of premium audio products. The region's market size was valued at $3.2 billion in 2019 [1].

3. **Asia Pacific:** The Asia Pacific region is witnessing robust growth in the headphone market, primarily due to an increase in smartphone users and rising disposable income levels. China and India are emerging as key markets in the region. The Asia Pacific headphone market size was valued at $5.5 billion in 2019 [1].

4. **Latin America:** Latin America is experiencing steady growth in the headphone market, driven by the growing popularity of portable audio devices and music streaming services. The region's market size was valued at $1.1 billion in 2019 [1].

5. **Middle East and Africa:** The Middle East and Africa region are also witnessing a rise in the demand for headphones, primarily driven by the growing youth population and increasing smartphone penetration. The region's market size was valued at $1.1 billion in 2019 [1].

Key Players

The global headphone market is highly competitive, with numerous players competing for market share. Some of the key players in the industry include:

- Apple Inc.
- Sony Corporation
- Samsung Electronics Co., Ltd.
- Bose Corporation

- Sennheiser Electronic GmbH & Co. KG
- JBL (Harman International Industries, Inc.)
- Skullcandy Inc.
- Beats Electronics LLC
- Plantronics, Inc.

These companies differentiate themselves through product innovation, brand reputation, and strategic partnerships. As technology continues to evolve, we can expect increased competition and new entrants in the market.

Future Outlook

The headphone market is projected to witness continued growth in the coming years. Key trends and factors that will shape the future of the market include:

1. **Wireless and true wireless earbuds:** The demand for wireless and true wireless earbuds is expected to surge as more smartphone manufacturers phase out the 3.5mm audio jack. These earbuds offer enhanced portability, convenience, and seamless integration with other devices.

2. **Noise-canceling technology:** Noise-canceling headphones are gaining popularity, driven by the need for immersive audio experiences, particularly in busy urban environments and during travel. Advancements in noise-canceling technology will further improve the user experience.

3. **Smart features and voice assistants:** Headphones are becoming smarter, with built-in voice assistants such as Siri, Google Assistant, and Alexa. These features offer hands-free controls, voice commands, and integration with smart devices.

4. **Health and fitness tracking:** With the increasing emphasis on health and wellness, the integration of health and fitness tracking features into headphones is becoming more common. Biometric sensors and heart rate monitors are examples of such features that cater to fitness enthusiasts.

5. **Augmented reality (AR) and virtual reality (VR) integration:** AR and VR technologies are expected to have a significant impact on the headphone market. Headphones with spatial audio capabilities and immersive sound reproduction will enhance the AR/VR experience.

In conclusion, the global headphone market continues to expand, driven by factors such as technological advancements, increasing smartphone penetration, and changing consumer preferences. With a projected growth rate of 3.1% from 2020 to 2027, this market offers significant opportunities for both established players and new entrants. As consumers seek enhanced audio experiences and new features, innovation will be a key differentiator in this highly competitive market.

References

[1] Grand View Research. (2020). Headphones Market Size, Share & Trends Analysis Report By Product (In-Ear, Over-Ear), By Price (Less Than 50, Between 50-100, Above 100), By Technology, By Application, By Region, And Segment Forecasts, 2020 - 2027.

Consumer Preferences and Purchasing Behaviors

Understanding consumer preferences and purchasing behaviors is essential in the study of marshmallows and headphones. This knowledge provides valuable insights into the demand for these products, the factors that influence consumer choices, and the trends that shape the market. In this section, we will explore the various aspects of consumer preferences and purchasing behaviors related to marshmallows and headphones.

Factors Influencing Consumer Preferences

Consumer preferences for marshmallows and headphones are influenced by a variety of factors. These factors can be broadly categorized into internal (individual) and external (environmental) factors.

Internal Factors Internal factors are personal characteristics that affect consumer preferences. These include:

- **Personal Taste:** Consumers' personal taste and preferences for flavors, textures, and sensory experiences play a significant role in their choice of marshmallows. Some individuals prefer traditional flavors like vanilla, while others may seek out unique and exotic flavors.

- **Cultural Background:** Cultural background and upbringing also influence consumer preferences. For example, certain cultures may have traditional desserts or dishes that incorporate marshmallows, leading to a stronger affinity for these products.

- **Lifestyle and Health Choices:** Lifestyle factors, such as dietary restrictions, health concerns, or specific dietary choices (e.g., vegan or gluten-free), can affect consumers' preference for marshmallows. Health-conscious consumers may opt for marshmallows with low sugar or natural ingredients.

- **Music Preferences:** In the case of headphones, consumers' music preferences and listening habits significantly impact their choice. Audiophiles may prioritize sound quality and select high-end, specialized headphones, while casual listeners may opt for more affordable options.

External Factors External factors refer to the environmental and social influences on consumer preferences. These include:

- **Marketing and Advertising:** Effective marketing campaigns and creative advertising can shape consumer preferences for marshmallows and headphones. Clever branding, endorsements by popular influencers, and engaging storytelling can sway consumer choices.

- **Social Influences:** The opinions and recommendations of friends, family, and online communities can significantly impact consumer preferences. Positive reviews, social media trends, and word-of-mouth recommendations may guide consumers' decisions.

- **Price and Value Perception:** Price is a crucial factor for many consumers. The perceived value of a product in relation to its price often determines whether consumers perceive it as worth purchasing. Factors such as affordability, quality, and durability influence consumers' perception of value.

- **Product Features and Innovation:** Consumer preferences are influenced by the features and innovations offered by marshmallow and headphone manufacturers. For example, headphone features like noise cancellation, wireless connectivity, and durability can make a product more appealing to consumers.

By considering both internal and external factors, marketers and manufacturers can better understand and cater to consumer preferences, thereby gaining a competitive advantage in the market.

CHAPTER 5: ECONOMIC AND ENVIRONMENTAL CONSIDERATIONS OF MARSHMALLOWS AND HEADPHONES

Consumer Purchasing Behaviors

Consumer purchasing behaviors encompass the actions individuals take while making buying decisions for marshmallows and headphones. Understanding these behaviors is crucial for businesses to develop effective marketing strategies and meet consumer demands. Several key aspects of consumer purchasing behaviors are worth considering.

Purchase Motivation Consumer purchase decisions for marshmallows and headphones are motivated by a range of factors. These include:

- **Functional Needs:** Consumers purchase marshmallows for their inherent flavor, sweetness, and versatility in a wide range of recipes. Headphones are primarily bought for personal audio enjoyment, audio production, or communication purposes.

- **Emotional Needs:** Emotional factors also play a role in customer decisions. Marshmallows can evoke feelings of nostalgia, comfort, and indulgence, while headphones can provide escapism, relaxation, or enhanced listening experiences.

- **Social Needs:** Consumers may also be motivated by social factors. For example, purchasing marshmallows for a camping trip or a social gathering can fulfill social and recreational needs. Buying headphones that align with current fashion or trends may enhance social acceptance or status.

Decision-Making Process Consumer purchasing decisions typically involve a series of steps known as the decision-making process. This process consists of the following stages:

1. **Problem Recognition:** Consumers recognize a need or desire for marshmallows or headphones, triggered by internal or external stimuli.

2. **Information Search:** Consumers seek information about marshmallow and headphone options through various sources, such as online reviews, social media, product descriptions, and recommendations.

3. **Evaluation of Alternatives:** Consumers compare and evaluate different marshmallow and headphone options based on factors like price, quality, brand reputation, features, and personal preferences.

4. **Purchase Decision:** Consumers make their final purchase decision, considering factors such as affordability, availability, convenience, and perceived value.

5. **Post-Purchase Evaluation:** After purchase, consumers evaluate their satisfaction with the chosen marshmallows or headphones based on their expectations and usage experience. Positive experiences can lead to brand loyalty and repeat purchases.

Shopping Channels Consumer preferences and purchasing behaviors are also influenced by the shopping channels available to them. These channels include traditional brick-and-mortar stores, e-commerce platforms, and direct-to-consumer channels.

- **Brick-and-Mortar Stores:** Physical stores offer consumers the opportunity to experience marshmallows and headphones firsthand, allowing them to touch, feel, and test the products before making a purchase. These stores provide in-person customer service and immediate product availability.

- **E-commerce Platforms:** Online shopping platforms provide convenience and a wide range of options. Consumers can compare prices, read reviews, and make purchases from the comfort of their homes. Additionally, e-commerce platforms enable sellers to reach a global market.

- **Direct-to-Consumer Channels:** Some manufacturers choose to sell their products directly to consumers through their websites or flagship stores. By eliminating intermediaries, these channels allow for a more personalized shopping experience, stronger brand-consumer relationships, and potential cost savings.

Consumer Trends Monitoring consumer trends is crucial for businesses to stay relevant and meet evolving consumer preferences. Some current trends in consumer preferences and purchasing behaviors for marshmallows and headphones include:

- **Health-conscious Choices:** Consumers are increasingly seeking healthier alternatives in both marshmallows and headphones. This includes marshmallows made with natural ingredients, low sugar options, and headphones with built-in health tracking features.

- **Sustainability and Ethical Considerations:** There is a growing emphasis on sustainability and ethical practices in consumer choices. Marshmallow and headphone manufacturers that demonstrate environmentally friendly production processes, fair trade sourcing, and recycling initiatives are more likely to attract environmentally conscious consumers.

- **Personalization and Customization:** Consumers appreciate personalized experiences and customized products. Marshmallow and headphone manufacturers that offer options for personalized flavors, packaging, or headphone design can cater to this trend.

- **Influence of Social Media and Online Communities:** Social media and online communities play a significant role in shaping consumer preferences and purchasing behaviors. Influencers and online recommendations influence product choices and drive trends, especially among younger demographics.

By understanding consumer preferences, purchasing behaviors, and keeping abreast of consumer trends, businesses can tailor their marketing strategies, product offerings, and customer experiences to meet the demands of the market.

Overall, studying consumer preferences and purchasing behaviors for marshmallows and headphones provides valuable insights into the diverse factors that influence consumer choices, the decision-making process, and the trends that shape the market. This knowledge is essential for businesses and marketers to create successful strategies, develop innovative products, and meet consumer demand in this globally interconnected world.

Environmental Impact of Headphone Manufacturing and Disposal

The manufacturing and disposal of headphones have significant environmental implications, ranging from the extraction of raw materials to the disposal of electronic waste. This section explores the key environmental considerations associated with headphone production and offers insights into potential solutions.

Extraction and Processing of Raw Materials

Headphones are comprised of various materials, including plastic, metal, rubber, and electronic components. The extraction and processing of these raw materials can have detrimental effects on the environment. For example, the extraction of metals like aluminum and copper often involves destructive mining practices that can lead to deforestation, habitat destruction, and water pollution. Additionally, the production of plastic components contributes to the growing issue of plastic waste and pollution.

To minimize these impacts, headphone manufacturers can adopt sustainable sourcing practices, such as using recycled materials or seeking out suppliers with responsible mining certifications. Additionally, the development of alternative materials, such as bio-based plastics or biodegradable materials, can reduce the reliance on fossil fuel-based plastics and minimize the environmental footprint of headphone production.

Energy Consumption and Emissions

The manufacturing process of headphones requires significant energy consumption, contributing to greenhouse gas emissions and climate change. The extraction and processing of raw materials, assembly, and transportation all require energy inputs, primarily from non-renewable sources like fossil fuels.

To address these concerns, headphone manufacturers can implement energy-efficient technologies and processes, such as optimizing production lines, utilizing renewable energy sources, and improving supply chain logistics to reduce transportation emissions. By adopting cleaner energy sources and implementing energy-saving measures, manufacturers can mitigate their carbon footprint and contribute to a more sustainable headphone industry.

Electronic Waste and Disposal

The disposal of electronic waste, including outdated or broken headphones, is a pressing environmental issue. Improper disposal can lead to the release of hazardous substances, such as heavy metals and toxic chemicals, into the environment, posing risks to both human health and ecosystems. Moreover, electronic waste often ends up in landfills, where it can contaminate soil and water.

To mitigate the environmental impact of headphone disposal, manufacturers can promote responsible e-waste management practices. This includes establishing and supporting recycling programs, implementing take-back initiatives, and designing headphones with modular components that can be easily repaired or upgraded. By encouraging consumers to recycle their old headphones and providing accessible recycling options, manufacturers can reduce the amount of electronic waste and promote a circular economy.

Product Design and Lifecycle Assessment

A sustainable approach to headphone manufacturing involves considering the entire product lifecycle, from design to disposal. Conducting a lifecycle assessment (LCA) enables manufacturers to evaluate and minimize the environmental impacts of their products. This includes assessing the environmental impacts of raw material extraction, manufacturing processes, energy consumption, transportation, and end-of-life disposal.

By incorporating eco-design principles, manufacturers can optimize their products for minimal environmental impact. This may involve using fewer materials, designing for disassembly and recyclability, reducing energy consumption during use, and improving durability to prolong product lifespan. Additionally, providing easily accessible information on the environmental footprint and recyclability of headphones can foster consumer awareness and encourage more sustainable purchasing decisions.

Case Study: Sustainable Headphone Production

To illustrate sustainable practices in headphone manufacturing, let's consider a case study of a headphone company that has implemented innovative environmental strategies.

Company X: Company X is a leading headphone manufacturer committed to sustainability. They have implemented several initiatives to minimize their environmental impact throughout the production process.

Initiative 1: Material Sourcing: Company X works closely with their suppliers to ensure responsible sourcing of raw materials. They prioritize suppliers with fair trade and responsible mining certifications, reducing the negative environmental and social impacts associated with material extraction.

Initiative 2: Energy Efficiency: Company X has invested in energy-efficient technologies and practices within their manufacturing facilities. They have installed solar panels on the premises, significantly reducing their reliance on non-renewable energy sources. Additionally, they continuously optimize their production processes to minimize energy consumption and waste generation.

Initiative 3: Recycling Program: Company X has established a comprehensive recycling program that encourages customers to return their old headphones for proper disposal. They partner with e-waste recycling facilities to ensure the safe and ethical handling of electronic waste. Moreover, they prioritize the use of modular components in their headphone design, allowing for easier repair and recycling.

Initiative 4: **Consumer Awareness:** Company X is committed to raising consumer awareness about the environmental impact of headphones. They provide educational materials on their website, highlighting the importance of responsible disposal and offering information on local recycling options. By engaging consumers in the sustainability conversation, they aim to foster more conscious purchasing decisions.

Through these initiatives, Company X demonstrates that sustainable headphone manufacturing is achievable by incorporating environmental considerations into all stages of the product lifecycle.

Conclusion

The environmental impact of headphone manufacturing and disposal is a significant concern in our modern society. By adopting sustainable practices, manufacturers can minimize the extraction of raw materials, reduce energy consumption, promote responsible e-waste management, and design products with a lower environmental footprint.

Consumers also play a crucial role in mitigating the impact of headphones on the environment. By choosing sustainably manufactured headphones, recycling their old devices, and supporting companies that prioritize environmental stewardship, individuals can contribute to a more sustainable headphone industry for future generations.

Corporate Social Responsibility in the Marshmallow and Headphone Industries

Ethical Sourcing and Fair Trade Practices

In today's globalized world, ethical sourcing and fair trade practices play an increasingly important role in the business landscape. This holds true for the marshmallow and headphone industries as well. Ethical sourcing refers to the procurement of raw materials, ingredients, or components from suppliers who uphold social and environmental standards, while fair trade practices ensure that workers are treated fairly and receive a fair wage for their labor. This section explores the significance of ethical sourcing and fair trade practices in these industries, the challenges they face, and the potential solutions to address them.

The Importance of Ethical Sourcing

Ethical sourcing is essential for sustainable and responsible business practices. It encompasses various aspects, including respect for workers' rights, environmental conservation, and community well-being. By engaging in ethical sourcing, companies can reduce their negative impact on the planet and society while fostering positive change. In the context of marshmallows and headphones, ethical sourcing becomes particularly relevant due to the reliance on agricultural resources, such as sugar and gelatin for marshmallows, and rare earth metals for headphones.

Challenges in Ethical Sourcing

Despite the importance of ethical sourcing, there are several challenges that companies in the marshmallow and headphone industries might face. One of the primary challenges is the complex and fragmented supply chain. Raw materials for marshmallows and headphones often come from different regions around the world, making it difficult to trace their origin and ensure ethical practices at every step. Additionally, the lack of transparency in some supply chains can lead to unethical labor practices, such as child labor or poor working conditions.

Another challenge is the economic pressure faced by suppliers and manufacturers. In order to meet market demand and stay competitive, some companies may compromise ethical standards to reduce costs. This includes sourcing materials from suppliers who do not adhere to fair labor practices or environmental regulations. Balancing profitability with ethical principles can be a delicate task for businesses.

Solutions and Best Practices

To address the challenges in ethical sourcing, companies can adopt various solutions and best practices. Firstly, establishing transparency in the supply chain is crucial. This can be achieved through rigorous auditing and certification processes, along with the use of technology such as blockchain to track and verify the origin of raw materials. By ensuring transparency, companies can identify potential ethical issues and take appropriate actions to rectify them.

Collaboration among industry stakeholders is another important solution. This includes sharing best practices, conducting joint audits, and supporting initiatives that promote ethical sourcing and fair trade. By working together, companies can leverage collective knowledge and resources to create positive change throughout the supply chain.

Implementing fair trade practices is also paramount. Companies can work towards obtaining fair trade certifications, which guarantee that the workers involved in the production process receive fair wages and work in safe conditions. Moreover, fostering long-term relationships with suppliers and providing capacity-building support can enhance ethical practices and drive positive social impact.

Real-World Examples and Initiatives

Several initiatives and organizations are actively working towards promoting ethical sourcing and fair trade practices in the marshmallow and headphone industries. For instance, the Fairtrade Foundation certifies certain marshmallow brands and encourages fair pricing and social sustainability. This ensures that farmers and workers involved in producing marshmallow ingredients receive fair compensation and are protected from exploitation.

In the headphone industry, organizations such as the Responsible Business Alliance (RBA) advocate for ethical practices in the supply chain. The RBA provides tools and resources for companies to address issues such as forced labor, working hours, and health and safety. By joining such initiatives, companies demonstrate their commitment to ethical sourcing and fair trade.

Furthermore, consumer awareness and demand for ethically sourced products have been instrumental in driving change. As consumers become more conscious of the social and environmental impact of their purchasing decisions, they are seeking products that align with their values. This trend encourages companies to prioritize ethical sourcing as a means of gaining a competitive edge and building a positive brand image.

Conclusion

Ethical sourcing and fair trade practices have become imperative in the marshmallow and headphone industries. By embracing transparency, collaboration, and fair trade certifications, companies can ensure that their products are produced in a socially and environmentally responsible manner. Ethical sourcing not only benefits the workers, communities, and the planet but also helps companies build trust, loyalty, and long-term sustainability. As consumers continue to prioritize ethical considerations, companies that champion ethical sourcing will be well-positioned for success in the global marketplace.

Recycling Initiatives and Sustainability Commitments

In recent years, there has been a growing concern about the environmental impact of industrial production and consumption. The marshmallow and headphone industries are no exception to this concern. As consumers become more conscious of the need to reduce waste and promote sustainable practices, companies in these industries have started implementing recycling initiatives and sustainability commitments. In this section, we will explore the various recycling and sustainability efforts undertaken by marshmallow and headphone manufacturers, as well as their environmental impacts and future prospects.

Recycling Initiatives in the Marshmallow Industry

The production and packaging of marshmallows generate a significant amount of waste, including plastic wrappers and containers. To address this issue, many marshmallow manufacturers have introduced recycling programs to collect and recycle their packaging materials. These programs aim to reduce the environmental impact of marshmallow consumption by minimizing the amount of waste going to landfills.

One example is the "Marshmallow Wrapper Recycling" initiative launched by a leading marshmallow brand. This program encourages consumers to collect empty marshmallow wrappers and send them back to the company for recycling. The wrappers are then processed and used to create new packaging materials or other plastic products. By engaging consumers in the recycling process, this initiative not only reduces waste but also raises awareness about the importance of responsible consumption.

Another innovative approach to recycling in the marshmallow industry is the development of edible packaging materials. Some companies are experimenting with using biodegradable and edible materials, such as edible films made from plant-based sources, as wrappers for marshmallows. These materials eliminate the need for traditional packaging and reduce waste. Moreover, they offer a unique and sustainable consumer experience, as consumers can enjoy the entire marshmallow, including the wrapper.

Sustainability Commitments in the Headphone Industry

The headphone industry faces sustainability challenges related to the production process, material sourcing, and end-of-life disposal of headphones. In response to these challenges, many headphone manufacturers have made sustainability commitments to reduce their environmental impact and promote responsible practices.

One major focus of sustainability commitments in the headphone industry is the use of recycled materials in headphone production. Companies are exploring ways to incorporate recycled plastics and metals into their manufacturing processes. For example, a leading headphone brand has introduced a line of headphones made from recycled ocean plastics, raising awareness about marine pollution and the importance of recycling.

In addition to material sourcing, headphone manufacturers are also addressing the issue of electronic waste through recycling programs. Many companies offer headphone recycling services, allowing consumers to return their old or broken headphones for proper disposal. These programs aim to recover valuable components and materials from headphones, reducing the need for virgin resources and minimizing the environmental impact of headphone disposal.

Furthermore, headphone manufacturers are exploring sustainable energy solutions for their wireless headphone models. This includes the use of rechargeable batteries and energy-efficient technologies to minimize power consumption. Some companies are even developing headphones that can be charged using renewable energy sources, such as solar power, providing users with an eco-friendly and sustainable listening experience.

Environmental Impact and Future Prospects

The recycling initiatives and sustainability commitments in the marshmallow and headphone industries have the potential to significantly reduce their environmental footprint. By promoting responsible consumption and waste management practices, these efforts contribute to the overall goal of achieving a circular economy, where resources are reused, recycled, or composted.

However, it is important to acknowledge the challenges and limitations of these initiatives. For marshmallow manufacturers, the development and implementation of edible packaging materials are still in the early stages, and further research and development are needed to ensure the safety, quality, and scalability of these materials. In the headphone industry, increasing the use of recycled materials depends on the availability and accessibility of such materials in the market.

Moreover, raising consumer awareness and changing consumption patterns remain crucial for the success of recycling initiatives and sustainability commitments. Education campaigns and incentives can play a vital role in encouraging consumers to actively participate in recycling programs and make environmentally conscious choices.

Looking ahead, advancements in material science and recycling technologies hold great potential for further improving the sustainability of the marshmallow and headphone industries. Continued collaboration between manufacturers, researchers, and policymakers will be essential in driving innovation and promoting sustainable practices. By embracing recycling initiatives and sustainability commitments, these industries can pave the way for a more environmentally friendly and socially responsible future.

Summary

In this section, we explored the recycling initiatives and sustainability commitments undertaken by the marshmallow and headphone industries. We discussed how marshmallow manufacturers have introduced recycling programs, such as wrapper collection initiatives and edible packaging materials, to reduce waste and promote responsible consumption. Similarly, we examined the sustainability commitments of headphone manufacturers, including the use of recycled materials, headphone recycling programs, and the development of energy-efficient and renewable-powered headphones.

While these efforts have the potential to significantly reduce the environmental impact of the marshmallow and headphone industries, challenges and limitations persist. Further research and development are needed for edible packaging materials in the marshmallow industry, and the availability of recycled materials in the market is crucial for the headphone industry. Additionally, consumer awareness and behavior change are essential for the success of recycling initiatives and sustainability commitments.

Looking forward, continued collaboration and innovation will play a vital role in improving the sustainability of these industries. Advancements in material science and recycling technologies offer promising prospects for a more environmentally friendly future. By embracing recycling initiatives and sustainability commitments, the marshmallow and headphone industries can contribute to a circular economy and pave the way for a more sustainable world.

Chapter 6: Technological Advances in Marshmallows and Headphones

Innovations in Marshmallow Manufacturing

New Ingredients and Flavor Enhancements

In recent years, there have been significant advancements in the field of marshmallow manufacturing, particularly in the area of ingredients and flavor enhancements. These developments have not only expanded the variety of marshmallow flavors available in the market but have also improved the overall taste and quality of marshmallows. In this section, we will explore some of the new ingredients and flavor enhancements that have revolutionized the marshmallow industry.

Innovative Ingredients

Traditionally, marshmallows were made using a combination of sugar, water, and gelatin. However, with the growing demand for unique and exotic flavors, manufacturers have started experimenting with new ingredients to create innovative marshmallow variations. Here, we highlight a few of these inventive ingredients:

- **Fruit Puree:** To add a burst of fruity flavor to marshmallows, fruit purees such as strawberry, raspberry, or mango are now being incorporated into the marshmallow mix. These natural purees not only enhance the taste but also contribute to the visual appeal of the marshmallows, giving them a vibrant color.

- **Herbs and Spices:** Marshmallows infused with herbs and spices have gained popularity among consumers seeking new and exciting taste experiences. Ingredients such as lavender, cinnamon, mint, and cardamom can elevate the flavor profile of marshmallows, making them more versatile and appealing to a wider audience.

- **Floral Extracts:** Floral extracts like rose, violet, and hibiscus bring a delicate and aromatic touch to marshmallows. These extracts not only impart unique flavors but also lend a sophisticated and elegant appeal to the final product. Floral marshmallows are often featured in gourmet dessert recipes or used as indulgent treats for special occasions.

These innovative ingredients not only provide a wider range of flavor options but also cater to evolving consumer preferences for natural and exciting taste experiences.

Enhanced Flavor Techniques

Apart from incorporating new ingredients, marshmallow manufacturers have also developed various techniques to enhance the flavor of marshmallows. These flavor enhancements can be achieved through different processes during production. Let's explore a few of these techniques:

- **Infusions and Steeping:** Infusing marshmallow mixtures with flavors from ingredients like tea leaves, coffee beans, or spices allows the flavors to meld together, creating a more complex and robust taste. Steeping these ingredients in hot water or milk before incorporating them into the marshmallow mixture helps extract the full flavor profile and imparts a rich taste to the final product.

- **Flavor Marbling:** Marshmallows can be visually stunning by incorporating flavor marbling techniques. This process involves layering or swirling different flavors or colors together to create visually appealing patterns. For example, chocolate and vanilla marshmallows can be marbled together to create a beautiful swirl effect, adding both visual interest and enhanced taste.

- **Coating and Dusting:** Marshmallows are often coated or dusted with various ingredients to add an extra layer of flavor. This can be done by coating the marshmallows in powdered sugar, cocoa, or flavored syrups, enhancing the sensory experience by providing additional taste and texture dimensions.

These flavor enhancement techniques have not only transformed the taste of marshmallows but have also opened up new possibilities for culinary creativity and experimentation.

Real-World Examples

To better understand the impact of new ingredients and flavor enhancements in the marshmallow industry, let's consider a real-world example.

One popular brand, Marshmallow Delights, has successfully introduced a range of innovative flavors using these advancements. They offer marshmallows infused with exotic fruit purees such as passionfruit and guava, providing a tropical twist to the classic marshmallow. By incorporating natural flavors, Marshmallow Delights has attracted a niche market of consumers looking for unique taste experiences.

Another example is Flavorsome Mallows, a company known for their visually appealing marbled marshmallows. They use a combination of traditional and innovative ingredients to create stunning patterns and flavors. Their blueberry and lavender marbled marshmallows have been a hit among consumers, combining the refreshing taste of blueberries with the aromatic and relaxing qualities of lavender.

These examples demonstrate how new ingredients and flavor enhancements have allowed marshmallow manufacturers to cater to diverse consumer preferences and create exciting products that go beyond the traditional marshmallow flavor.

Caveats and Considerations

While the introduction of new ingredients and flavor enhancements has expanded the possibilities in the marshmallow industry, there are a few caveats and considerations to keep in mind:

- **Allergies and Dietary Restrictions:** Manufacturers must be cautious when introducing new ingredients to avoid potential allergens or ingredients that may not be suitable for individuals with certain dietary restrictions. Clear labeling and comprehensive ingredient lists are essential to ensure consumer safety and transparency.

- **Quality Control and Consistency:** With the introduction of new ingredients and flavor techniques, it is crucial for manufacturers to maintain quality control and consistency. Consumer expectations for taste and texture should be met consistently across different batches and production runs.

- **Market Acceptance and Demand:** While innovation is important, it is equally important to gauge market acceptance and demand for new flavors. Conducting market research and gathering feedback from consumers can help manufacturers understand which flavors are most likely to succeed in the market.

By being mindful of these caveats and considerations, marshmallow manufacturers can continue to push the boundaries of flavor and create exciting products that cater to evolving consumer tastes.

Conclusion

The development of new ingredients and flavor enhancements has brought about a paradigm shift in the marshmallow industry. Through the incorporation of innovative ingredients and the application of flavor enhancement techniques, manufacturers have successfully expanded the variety of marshmallow flavors and improved the overall taste and quality of marshmallows. These advancements have not only satisfied the cravings of consumers seeking unique and exciting taste experiences but have also provided opportunities for culinary creativity and experimentation. However, it is crucial for manufacturers to strike a balance between innovation and consumer preferences, ensuring that flavors are not only enticing but also safe and suitable for individuals with different dietary needs. By embracing new ingredients and flavor enhancements while considering the caveats and considerations, marshmallow manufacturers can continue to captivate consumers and drive the growth of this vibrant industry.

Production Automation and Efficiency

In today's fast-paced and competitive marshmallow industry, production automation and efficiency play a crucial role in meeting the growing demand for marshmallows worldwide. The use of advanced technology and automation systems allows manufacturers to streamline their production processes, increase output, and reduce costs. In this section, we will explore the various aspects of production automation and efficiency in the marshmallow industry.

Benefits of Production Automation

Production automation offers several benefits for marshmallow manufacturers. Firstly, it significantly reduces labor requirements. Tasks that were previously carried out manually, such as mixing ingredients, molding, and packaging, can now be efficiently automated. This not only saves time but also reduces the risk of human error, ensuring consistent quality throughout the production process.

Secondly, automation improves production speed. Automated machinery can produce marshmallows at a much faster rate compared to manual labor. This is particularly beneficial when dealing with large-scale production volumes or during peak seasons when demand is high. Increased production speed allows

INNOVATIONS IN MARSHMALLOW MANUFACTURING

manufacturers to meet customer demand more effectively and maintain a competitive edge in the market.

Furthermore, automation systems enhance product consistency and quality control. By automating critical production steps, manufacturers can ensure that each marshmallow meets the desired specifications in terms of size, shape, texture, and taste. Consistency in product quality helps build customer trust and loyalty, leading to increased sales and market share.

Automated Marshmallow Production Line

An automated marshmallow production line typically consists of several interconnected machines that handle different stages of the manufacturing process. Let's take a closer look at some of the key components of an automated marshmallow production line.

| Mixing and Cooking | → | Molding |

Figure 0.1: Mixing and cooking stage in an automated marshmallow production line.

Mixing and Cooking The mixing and cooking stage is where the ingredients are combined and cooked to create the marshmallow base. A mixer, typically a large-scale industrial mixer, blends the ingredients, including sugar, corn syrup, gelatin, and flavorings, to form a homogenous mixture. The mixer ensures that all ingredients are evenly distributed, resulting in consistent taste and texture.

The mixed ingredients are then transferred to a cooker, where they are heated and cooked to the desired temperature and texture. The cooker ensures that the marshmallow base reaches the correct consistency and viscosity before proceeding to the next stage. Automated cooking systems can precisely control the cooking parameters, such as temperature and cooking time, to achieve the desired product characteristics.

Molding Once the marshmallow base is cooked, it is ready to be shaped and molded. In an automated production line, the marshmallow mixture is pumped from the cooker to a molding machine. The molding machine consists of a series of molds or nozzles that shape the marshmallow mixture into desired forms, such as cubes, cylinders, or novelty shapes.

The molding machine operates with high precision, ensuring consistent sizing and shape for each marshmallow. The automated process reduces the need for manual intervention, minimizing the risk of product inconsistencies and defects.

Drying and Packaging After molding, the marshmallows need to be dried to remove excess moisture and obtain the desired texture. In an automated production line, the molded marshmallows are transferred to a drying conveyor. The conveyor transports the marshmallows through a temperature-controlled drying tunnel, where warm air gently removes moisture without affecting the product quality.

Once the marshmallows have been dried, they are ready for packaging. Automated packaging machines can handle various packaging formats, such as bags, boxes, or individual wrappers. These machines efficiently seal the marshmallows, ensuring that they remain fresh and protected during storage and transportation.

Challenges and Considerations

While production automation offers numerous benefits, it also presents some challenges that manufacturers need to consider.

One challenge is the initial investment cost. Implementing an automated production line requires significant capital investment in machinery, robotics, and control systems. Manufacturers need to carefully assess the expected return on investment and weigh it against the potential long-term benefits.

Another consideration is the need for skilled technicians and engineers to operate and maintain the automated systems. Proper training and expertise are essential to ensure the smooth functioning of the production line, minimize downtime, and address any technical issues that may arise.

Manufacturers must also address food safety and hygiene requirements, as well as comply with regulatory standards. Automation systems need to be designed and operated in a way that ensures product safety and minimizes the risk of contamination. Regular cleaning, sanitization, and maintenance of the machinery are crucial for maintaining high-quality standards.

Case Study: Industrial Robots in Marshmallow Production

To illustrate the benefits of production automation in the marshmallow industry, let's consider a case study on the implementation of industrial robots.

Problem A marshmallow manufacturer wants to increase production efficiency and reduce labor costs. The manual process of transferring marshmallows from the molding machine to the drying conveyor is time-consuming and requires a significant workforce. The manufacturer is looking for a solution to automate this stage of the process.

Solution The manufacturer decides to introduce industrial robots into the production line to automate the marshmallow transfer process. The robots are programmed to pick up the molded marshmallows from the molding machine and place them onto the drying conveyor. The robotic system is equipped with vision sensors to ensure accurate picking and placing of the marshmallows.

Results By implementing industrial robots, the marshmallow manufacturer achieves several benefits. The production speed increases significantly, as the robots can handle a large number of marshmallows in a short period. The automation also reduces the need for manual labor, resulting in cost savings and improved efficiency.

Furthermore, the introduction of robots minimizes the risk of product damage and contamination caused by human handling. The robots ensure consistent spacing and alignment of the marshmallows on the drying conveyor, improving product quality and reducing waste.

Considerations While the introduction of industrial robots streamlines the marshmallow transfer process, the manufacturer needs to address several considerations. The robots must be properly programmed and calibrated to handle different marshmallow sizes and shapes. Regular maintenance and calibration of the robotic system are necessary to ensure optimal performance and prevent production disruptions.

Additionally, proper safety measures need to be in place to protect human workers interacting with the robots. Clear guidelines and training programs should be established to ensure the safe operation of the robots and minimize the risk of accidents or injuries.

Conclusion

Production automation and efficiency are key factors in the modern marshmallow industry. Implementing automated systems not only reduces labor requirements but also improves production speed, product consistency, and quality control.

Although there are initial investment costs and considerations, the long-term benefits are substantial.

In this section, we explored the benefits of production automation, the components of an automated marshmallow production line, and the challenges and considerations manufacturers need to address. We also examined a case study on the use of industrial robots in marshmallow production, illustrating the positive impact of automation on efficiency and product quality.

By embracing production automation and continually incorporating advancements in technology, the marshmallow industry can meet the growing demand for marshmallows, enhance product quality, and maintain a competitive edge in the global market.

Nanotechnology and Texture Modification

Nanotechnology, a branch of science and technology that deals with materials and devices at the nanoscale, has revolutionized many fields, including the food industry. In the context of marshmallows, nanotechnology offers exciting possibilities for texture modification and enhancement. By manipulating and engineering the properties of materials at the nanoscale, it is possible to create marshmallows with unique and improved textures.

Understanding Nanoscale Materials

To appreciate the impact of nanotechnology on texture modification, it is important to understand the behavior of materials at the nanoscale. At this scale, the properties of materials can significantly differ from their macroscopic counterparts. This is due to the increased surface area-to-volume ratio and the quantum mechanical effects that come into play.

One example of nanoscale materials relevant to texture modification in marshmallows is nanoparticles. These are particles with sizes ranging from 1 to 100 nanometers. They can be engineered to possess specific properties such as increased surface area, improved stability, and enhanced reactivity.

Nanoparticles for Texture Modification

Nanoparticles can be utilized in the production of marshmallows to modify their texture in various ways. Here are a few examples:

1. **Improving Stability and Shelf Life:**

Incorporating nanoparticles with antimicrobial properties, such as silver nanoparticles, can help extend the shelf life of marshmallows. These nanoparticles can inhibit the growth of microorganisms, preventing spoilage and maintaining the marshmallows' freshness.

2. **Controlling Foam Structure:**

   The texture of marshmallows primarily depends on the structure and stability of the foam within them. Nanoparticles can be used to modify the foam structure, resulting in marshmallows with desired textures. For example, silica nanoparticles can be employed to control the size and uniformity of the air bubbles within the marshmallow, leading to a more uniform and stable foam structure.

3. **Enhancing Creaminess:**

   Creaminess is an important sensory attribute in marshmallows. Nanoparticles can be used to enhance the creaminess by improving the suspension of fat droplets within the marshmallow matrix. By coating the fat droplets with nanoparticles, their aggregation can be minimized, leading to a smoother and more indulgent texture.

4. **Creating Novel Textures:**

   Nanoparticles, such as cellulose nanoparticles, can be incorporated into marshmallows to create novel textures. These nanoparticles can modify the rheological properties of the marshmallow matrix, resulting in unique and desirable textural sensations upon consumption.

Challenges and Considerations

While nanotechnology offers exciting possibilities for texture modification in marshmallows, there are several challenges and considerations to address:

1. **Regulatory Approval:**

   The use of nanoparticles in food products requires careful consideration of safety and regulatory approval. It is crucial to ensure that the nanoparticles used are non-toxic and do not pose any health risks.

2. **Uniform Distribution:**

   Achieving a uniform distribution of nanoparticles within the marshmallow matrix is essential to ensure consistent texture modification. Proper processing techniques and formulation strategies need to be developed to achieve this distribution.

3. **Consumer Acceptance:**

   Introducing nanotechnology-modified marshmallows to the market requires consumer acceptance. Communicating the benefits and safety of these innovations to consumers is vital for their adoption.

Case Study: Nano-Structured Marshmallow Foam

As a case study in the application of nanotechnology for texture modification in marshmallows, researchers at the University of Food Science have developed a nano-structured marshmallow foam using cellulose nanofibers.

Cellulose nanofibers, obtained from natural sources such as plant fibers, are biodegradable and have excellent mechanical properties. These nanofibers were incorporated into the marshmallow foam to enhance its texture and stability.

The cellulose nanofibers acted as structural reinforcements within the foam, improving its elasticity and reducing collapse during processing and storage. The resulting marshmallow foam exhibited a more resilient and fluffy texture compared to traditional marshmallows.

This nano-structured marshmallow foam also showed increased resistance to moisture absorption, leading to a longer shelf life. Additionally, sensory evaluations indicated that the nanotechnology-modified marshmallows were well received by consumers, highlighting the potential for commercialization of such products.

Conclusion

Nanotechnology holds immense potential for texture modification in marshmallows. Through the use of nanoparticles and nanoscale materials, it is possible to create marshmallows with improved stability, controlled foam structure, enhanced creaminess, and novel textures. However, challenges such as regulatory approval, distribution uniformity, and consumer acceptance need to be addressed. The case study of nano-structured marshmallow foam demonstrates the successful application of nanotechnology in creating marshmallows with unique and desirable textures. With further research and development, nanotechnology can continue to advance the field of marshmallow texture modification, providing consumers with innovative and enjoyable culinary experiences.

Headphone Engineering and Design

Driver Technologies and Sound Quality

Driver technologies play a crucial role in the sound quality and overall performance of headphones. The driver is the component responsible for converting electrical signals into sound waves. In this section, we will explore different types of driver technologies used in headphones and their impact on sound quality.

Dynamic Drivers

Dynamic drivers are the most common type of drivers found in headphones. They consist of a diaphragm, voice coil, and magnet. When an electrical signal is passed through the voice coil, it creates a magnetic field that interacts with the permanent magnet, causing the diaphragm to vibrate and produce sound waves.

One key advantage of dynamic drivers is their ability to reproduce a wide range of frequencies. They excel in producing deep bass tones and can handle high power levels. However, dynamic drivers may struggle to reproduce high-frequency details with the same level of accuracy and precision as other driver types.

Balanced Armature Drivers

Balanced armature drivers are a smaller and more lightweight alternative to dynamic drivers. They consist of a tiny armature suspended between two magnets. When an electrical signal is applied, the armature moves, resulting in sound production.

Balanced armature drivers are known for their accuracy and detail in sound reproduction, especially in the mid to high-frequency range. They are commonly used in in-ear monitors (IEMs) and provide excellent noise isolation. However, due to their smaller size, they may not produce the same level of bass response as dynamic drivers.

Planar Magnetic Drivers

Planar magnetic drivers utilize a thin and lightweight diaphragm with an embedded wire coil. The diaphragm is suspended between two magnetic arrays that generate a magnetic field. When an electrical signal passes through the wire coil, it interacts with the magnetic field, causing the diaphragm to vibrate and produce sound.

These drivers are known for their exceptional accuracy, speed, and low distortion. Planar magnetic headphones can reproduce a wide frequency range with great detail and clarity. They excel in delivering a balanced, natural sound signature. However, planar magnetic drivers are typically more expensive and bulkier compared to other driver types.

Electrostatic Drivers

Electrostatic drivers use a thin, electrically charged diaphragm placed between two stators. The diaphragm is coated with a conductive material and is held in place by a fixed perforated plate. When an audio signal is applied to the stators, an electric field is created that moves the diaphragm, producing sound.

Electrostatic drivers offer unparalleled levels of clarity, detail, and speed. They can reproduce extremely subtle nuances and deliver an incredibly realistic sound reproduction. However, electrostatic headphones require a specialized amplifier to drive them properly, and they tend to be more expensive and less common.

Hybrid Driver Systems

Some headphones utilize hybrid driver systems, combining different driver technologies to achieve a desired sound signature. For example, a common hybrid configuration is a combination of a dynamic driver for bass response and balanced armature drivers for greater detail in the mid and high frequencies.

By using a hybrid driver system, manufacturers can take advantage of the strengths of different driver types, aiming to provide a well-rounded and balanced sound experience.

Evaluating Sound Quality

The sound quality of headphones depends not only on the driver technology but also on various other factors such as enclosure design, frequency response, and impedance.

Frequency response is particularly important, as it determines how accurately headphones reproduce different frequencies. An ideal frequency response is one that is flat, meaning all frequencies are reproduced equally without any emphasis or attenuation.

Impedance is another critical factor to consider, as it affects how efficiently headphones convert electrical signals into sound. Headphones with high impedance may require more amplification to produce adequate volume levels.

When evaluating sound quality, it is essential to consider personal preferences. Some listeners may prefer a bass-heavy sound, while others may prioritize accurate reproduction of all frequencies. Ultimately, the best sound quality is subjective and depends on individual taste.

Case Study: Sennheiser HD800

To illustrate the impact of driver technology on sound quality, let's consider the Sennheiser HD800 headphones. These open-back headphones feature a unique ring radiator driver design, which combines elements from dynamic and balanced armature drivers.

The ring radiator driver of the HD800 allows for a large soundstage and exceptional detail retrieval. It delivers a balanced, transparent sound with excellent imaging capabilities. The driver's design minimizes resonance and distortion, providing a highly accurate and detailed listening experience.

The case study of the Sennheiser HD800 demonstrates how driver technology can significantly contribute to the overall sound quality and listening pleasure offered by headphones.

Conclusion

Driver technologies play a vital role in determining the sound quality and performance of headphones. Dynamic drivers are the most common type, offering a wide frequency range and powerful bass response. Balanced armature drivers excel in accuracy and detail, especially in the mid to high frequencies. Planar magnetic drivers provide exceptional clarity and low distortion. Electrostatic drivers offer unparalleled levels of realism and detail, albeit at a higher price point. Hybrid driver systems combine different technologies to achieve a balanced sound signature. Evaluating sound quality involves considering factors such as frequency response and impedance. Ultimately, the best sound quality is subjective and depends on personal preferences.

Comfort and Ergonomics in Headphone Design

In the field of headphone design, achieving optimal comfort and ergonomics is essential to provide users with a satisfying listening experience. Headphones that are uncomfortable or poorly designed can lead to fatigue, discomfort, and even long-term issues such as ear or head pain. In this section, we will explore the various factors that contribute to comfort and ergonomics in headphone design, and discuss innovative solutions to enhance user comfort.

Understanding Human Anatomy and Physiology

To design headphones that are comfortable to wear for extended periods, it is crucial to understand the anatomy and physiology of the human head and ears. The shape and size of the human head vary greatly between individuals, making it challenging to create a one-size-fits-all design.

The human ear is particularly sensitive and delicate, composed of several intricate structures responsible for transmitting sound signals to the brain. The outer ear, consisting of the pinna and ear canal, plays a role in sound localization and helps in capturing sound waves. The middle ear, including the eardrum and three tiny bones (ossicles), transmits sound vibrations to the inner ear. The inner ear, housing the cochlea, converts sound vibrations into electrical signals that the brain can interpret.

Pressure and Weight Distribution

One of the main contributors to headphone discomfort is pressure exerted by the ear cups against the head and ears. Excessive pressure can cause pain and discomfort, especially when applied to sensitive areas such as the pinna or the temporal region. To address this issue, headphone manufacturers utilize techniques to distribute pressure more evenly across the head.

One approach is the use of adjustable headbands that allow users to customize the fit according to their head size. Padded headbands can also provide additional cushioning and reduce pressure points. Ear cups with soft cushioning made from materials like memory foam can further enhance comfort and ensure a snug fit without excessive pressure.

Weight distribution is another important consideration in designing comfortable headphones. Heavy headphones can strain the neck muscles and cause discomfort, especially during prolonged use. Manufacturers now employ lightweight materials and innovative designs to minimize the overall weight of the headphones while maintaining structural integrity and audio quality.

Ear Pad Materials and Design

The design and materials used for ear pads significantly influence comfort and ergonomics. Ear pads should be soft, breathable, and hypoallergenic to prevent discomfort, sweating, and skin irritations. Additionally, they should provide adequate insulation to minimize sound leakage and ensure optimal audio quality.

Memory foam ear pads have gained popularity due to their ability to conform to the shape of the user's ears, providing a comfortable and customized fit. These ear pads distribute pressure evenly and relieve strain on the earlobes and surrounding areas. Fabric coverings or synthetic leather can be used to enhance comfort and durability while maintaining a luxurious feel.

Some headphones incorporate hybrid ear pads that combine different materials to address specific comfort and noise isolation requirements. For example, a headphone may use memory foam for the part that comes in contact with the ear, while employing a leather-like material for the outer layer to enhance noise isolation and durability.

Adjustable Fit and Clamping Force

To cater to a diverse range of head sizes and preferences, headphones often feature adjustable mechanisms for fit and clamping force. The clamping force refers to the pressure exerted by the headphone's headband on the user's head.

An adjustable fit allows users to customize the position of the ear cups, ensuring that they align comfortably with their ears. This feature is particularly beneficial for individuals with larger or smaller heads. Additionally, headphones with a swivel or pivot mechanism at the ear cups are effective in adapting to various head shapes and positions, accommodating different listening scenarios.

Proper clamping force is crucial for a secure fit without excessive pressure. Too loose a fit can cause headphones to slip off, while excessive clamping force can lead to discomfort and fatigue. Manufacturers employ various techniques, including spring-loaded headbands and ergonomic designs, to achieve an optimal balance of fit and clamping force.

Heat Dissipation and Sweat Resistance

During extended listening sessions, heat can build up between the user's ears and the ear cups, leading to discomfort and sweaty ears. To mitigate this issue, headphone manufacturers incorporate design features that promote heat dissipation and sweat resistance.

Ventilation channels or perforations in the ear cups allow air to circulate, preventing the hot, humid environment that can cause discomfort. Moisture-wicking materials used in the ear pads and headbands effectively reduce sweat accumulation and maintain a comfortable listening experience.

Unconventional Solution: Biometric Data Integration

Advances in wearable technology open up new possibilities for optimizing comfort and ergonomics in headphone design. By integrating biometric sensors into headphones, it becomes possible to monitor factors such as heart rate, temperature, and perspiration levels. This data can be used to dynamically adjust the fit, clamping force, and other parameters of the headphones in real-time, enhancing comfort based on the user's physiological responses.

For example, if a user's body temperature increases during physical activity while wearing headphones, the biometric sensors can detect this change and automatically adjust the fit to allow better heat dissipation. Similarly, if excessive sweating is detected, moisture-wicking materials in the ear pads can activate to enhance sweat resistance and maintain comfort.

This integration of biometric data can provide a personalized and adaptive listening experience, ensuring optimal comfort under varying conditions and minimizing potential discomfort caused by prolonged headphone use.

Summary

Comfort and ergonomics play a vital role in headphone design, as they directly impact the overall listening experience and user satisfaction. Understanding human anatomy, pressure distribution, adjustable fit, ear pad design, and heat dissipation are key factors in achieving optimal comfort. Biometric data integration presents an exciting opportunity to further enhance comfort by dynamically adapting headphone parameters based on the user's physiological responses. By prioritizing comfort and ergonomics, headphone manufacturers can create products that can be enjoyed for extended periods without causing discomfort or fatigue.

Wireless and Connectivity Innovations

In recent years, wireless and connectivity innovations have revolutionized the world of headphones. With advancements in technology, users can now enjoy the freedom of wire-free audio experiences. This section explores the various wireless and connectivity innovations that have transformed the headphone industry.

Wireless headphones have become increasingly popular due to their convenience and flexibility. Instead of being tethered to a device, users can now enjoy seamless audio streaming without the hassle of wires. Let's delve into some of the key wireless and connectivity technologies that have made this possible.

Bluetooth Technology

One of the most widely adopted wireless technologies in the headphone industry is Bluetooth. Bluetooth technology allows for short-range, wireless communication between devices such as smartphones, laptops, and headphones.

Bluetooth headphones use radio waves to transmit audio signals from the source device to the headphones. The major advantage of Bluetooth technology is its compatibility with a wide range of devices. Whether you're listening to music on your smartphone, tablet, or computer, Bluetooth headphones offer a universal connectivity solution.

Over the years, Bluetooth technology has evolved to provide better data transfer rates, improved audio quality, and reduced power consumption. The latest Bluetooth versions, such as Bluetooth 5.0, offer faster and more stable connections, allowing for high-quality audio streaming with minimal latency.

Wireless Range and Stability

While Bluetooth provides convenience, the wireless range and stability of headphones have been areas of concern. However, recent advancements have addressed these issues, enhancing the overall user experience.

The range of Bluetooth headphones has significantly improved with the introduction of Bluetooth 5.0. This technology allows for connectivity up to 800 feet in open spaces, making it ideal for both indoor and outdoor use. Users can move around freely without worrying about signal interruptions or audio quality degradation.

To ensure a stable connection, modern Bluetooth headphones utilize advanced features like dual-mode Bluetooth, which combines classic Bluetooth with low-energy Bluetooth. This helps to maintain a reliable connection while maximizing battery life. Additionally, innovative antenna designs and signal processing algorithms have further enhanced the stability of wireless connections.

Wireless Charging

Another exciting innovation in wireless headphones is the advent of wireless charging technology. With traditional wired headphones, users often face the hassle of untangling wires and dealing with cable-related issues. Wireless charging eliminates these inconveniences by allowing users to charge their headphones without using any cables.

Wireless charging relies on electromagnetic induction to transfer power from a charging pad or dock to the headphones. This technology offers an effortless and clutter-free charging experience. Simply placing the headphones on a compatible charging pad initiates the charging process.

Furthermore, wireless charging is becoming more prevalent in other electronic devices, such as smartphones and smartwatches. This trend suggests a future where various devices can share the same charging pads, making it even more convenient for users.

Smart Features and Digital Assistants

Wireless and connectivity innovations have also enabled headphones to offer smart features and integrate with digital assistants. Many modern headphones have built-in touch controls or buttons that allow users to easily manage their audio playback, adjust volume, and answer calls without reaching for their connected devices.

Moreover, popular digital assistants like Apple's Siri, Amazon's Alexa, and Google Assistant can be integrated with wireless headphones. Users can activate these assistants with voice commands, and they can provide real-time information, manage daily tasks, and even control smart home devices.

The integration of digital assistants with wireless headphones enhances the overall user experience by offering a hands-free and efficient way to interact with devices and perform various tasks.

Noise-Canceling and Ambient Sound Modes

Wireless headphones have also seen significant advancements in noise-canceling technology. Active noise-canceling (ANC) headphones use built-in microphones to analyze ambient sounds and produce anti-noise signals that cancel out background noise, providing a more immersive and focused listening experience.

Additionally, some wireless headphones offer ambient sound modes, which allow users to selectively hear their surroundings while still enjoying their music. This feature is particularly useful in situations where it is essential to stay aware of the environment, such as during outdoor activities or commuting.

The combination of wireless connectivity and noise-canceling technology has revolutionized the audio experience, allowing users to enjoy high-quality, uninterrupted sound without being tethered to their devices.

Conclusion

Wireless and connectivity innovations have transformed the headphone industry, offering users a new level of freedom and convenience. Technologies like Bluetooth, wireless charging, smart features, and noise-canceling have revolutionized the way we listen to audio.

As wireless and connectivity technologies continue to evolve, we can expect even more exciting advancements in the future. From improved battery life to enhanced audio quality, wireless headphones are poised to provide an unparalleled listening experience. With these innovations, the world of headphones has become more immersive, accessible, and enjoyable for everyone.

Crossroads of Marshmallows and Headphones: Edible Audio Technology

Edible Sound Transmitters and Receivers

In recent years, the intersection of technology and food has led to some truly innovative and remarkable advancements. One such development is the emergence of edible sound transmitters and receivers. These edible devices have the ability to transmit and receive sound signals, opening up new possibilities in the field of edible audio technology.

Background

The concept of edible sound transmitters and receivers is built upon the principles of wireless communication and the properties of sound waves. In wireless communication, signals are sent through the air or other mediums using various technologies such as radio waves, microwaves, and infrared. Sound waves, on the other hand, are mechanical waves that require a medium, such as air, water, or a solid, to propagate.

Traditionally, sound has been transmitted through the air using speakers or headphones, which convert electrical signals into vibrations that travel through the air and reach our ears. However, the idea of transmitting sound through edible devices takes this concept to a whole new level, offering unique opportunities and applications.

Principles of Edible Sound Transmitters and Receivers

Edible sound transmitters and receivers operate by utilizing edible materials that have the ability to transmit and receive sound waves. These materials are carefully selected and engineered to ensure they can effectively carry sound signals without compromising their safety and digestibility.

One principle used in edible sound transmission is the conduction of vibrations through solid materials. Edible devices can be designed to vibrate in response to incoming sound waves, and these vibrations can then be detected by specialized sensors or receivers. The vibration patterns can be converted back into electrical signals, which can be further amplified and converted into audible sound.

To achieve effective sound reception, edible devices utilize the principles of acoustics and resonance. By carefully shaping and designing the edible material, it is possible to enhance the reception of sound waves and improve the overall audio quality. Additionally, the placement and arrangement of sensors or receivers within the edible device play a crucial role in capturing and converting the sound signals.

Applications of Edible Sound Transmitters and Receivers

The emergence of edible sound transmitters and receivers presents exciting opportunities across various fields. Here are a few potential applications:

1. Enhancing the Dining Experience: Edible sound transmitters and receivers can be incorporated into dining experiences to add a multisensory element. For example, a restaurant could serve a dish with an edible transmitter that plays a matching sound when consumed, enhancing the perception of flavors and creating a unique dining experience.

2. Medical and Therapeutic Uses: In the medical field, edible sound transmitters and receivers could be used for targeted drug delivery. By embedding the transmitter within an edible capsule, medication could be delivered to specific areas of the digestive tract and triggered by sound signals. This could revolutionize the field of personalized medicine.

3. Entertainment and Gaming: Edible sound devices could be utilized in interactive entertainment experiences, where users consume edible devices that transmit audio cues or feedback. This could enhance the immersive nature of gaming and virtual reality applications.

4. Educational Tools: Edible sound transmitters and receivers could be employed in educational settings as innovative teaching tools. Students could consume edible devices that transmit educational audio content, turning the act of learning into a multisensory experience.

Challenges and Considerations

The development and implementation of edible sound transmitters and receivers come with several challenges and considerations. Here are a few key ones:

1. Safety and Digestibility: The edible materials used in these devices must meet rigorous safety standards to ensure they are safe for consumption. The materials should also be easily digestible to prevent any potential health risks.

2. Signal Range and Quality: Ensuring reliable transmission and reception of sound signals is essential. Designing edible devices that can transmit and receive sound over a sufficient range and maintain high-quality audio is a significant challenge.

3. Power Source: Providing a stable power source for edible devices presents a unique challenge. Battery-free designs or utilizing the body's own energy may be potential solutions.

4. Cost and Scalability: The cost of producing edible sound devices and their scalability for mass production are important considerations. The materials used should be cost-effective, and the manufacturing process should be scalable to meet market demand.

Conclusion

Edible sound transmitters and receivers represent a fascinating fusion of technology and gastronomy. Their potential applications across various industries open up new frontiers in the field of edible audio technology. While there are challenges to overcome, the possibilities for enhancing dining experiences, medical applications, entertainment, and education are truly exciting. The future of edible sound devices holds promise for a truly multisensory world.

Applications of Edible Audio Technology

The intersection of edible technology and audio technology has paved the way for various innovative applications that enhance our sensory experience and revolutionize the way we enjoy and interact with sound. In this section, we will explore some of the exciting applications of edible audio technology, highlighting their potential benefits and discussing their impact on various aspects of our lives.

Enhanced Dining Experiences

One of the most intriguing applications of edible audio technology is its integration into the realm of gastronomy. Imagine attending a fine dining experience where each course is not only a feast for the palate but also a symphony for the ears. Edible audio technology allows for the creation of interactive and immersive dining experiences by incorporating sound elements into the food itself.

For instance, using specially designed sound-emitting devices, chefs can synchronize the audio with the presentation of each dish, enhancing the sensory experience. These devices can be embedded within the edible components of a dish, such as a cake or a candy, and emit audio signals that complement the flavors and textures. The sounds can range from gentle melodies to dynamic compositions, reflecting the mood and theme of the dish. By engaging multiple senses simultaneously, these edible audio experiences elevate the enjoyment of food and create lasting memories for diners.

Audio-enhanced Confectionery

Edible audio technology also finds its application in the realm of confectionery, where it adds an extra layer of interactivity and entertainment to sweet treats. Imagine biting into a marshmallow and being greeted with a burst of music or sound effects. Edible audio devices can be embedded within confectionery products, such as lollipops or gumdrops, to surprise and delight consumers.

These audio devices can be programmed to play a wide range of sounds, from simple tunes to complex melodies. They can also be synchronized with specific actions, such as biting or chewing, to create dynamic and interactive soundscapes. This technology not only enhances the sensory experience of consuming confectionery but also encourages creativity and playfulness.

Audio-guided Culinary Education

Edible audio technology can also be utilized as an educational tool in the field of culinary arts. By incorporating audio components into cooking lessons and recipe books, aspiring chefs can receive audio-guided instructions that enhance their learning experience.

For example, a recipe book equipped with edible audio devices can provide step-by-step instructions, guiding the reader through the cooking process. As the reader progresses through the recipe, the audio component can provide real-time feedback and suggestions to ensure the desired outcome. This audio guidance not only enhances the learning experience but also caters to different learning styles, making culinary education more accessible and engaging.

Edible Audio Therapy

In addition to entertainment and education, edible audio technology has shown promise in the field of therapy and wellness. Sound therapy has long been used to promote relaxation and reduce stress. Edible audio technology takes this concept further by allowing individuals to consume sound as a form of therapy.

Specially designed edible audio devices can emit soothing sounds, such as calming melodies or nature sounds, which can be ingested as part of a therapeutic regimen. These edible audio therapy devices have the potential to provide a unique and portable approach to stress relief and relaxation. The act of consuming the sound adds a tactile element to the therapy, enhancing the overall experience and potentially increasing its effectiveness.

Caveats and Considerations

While the applications of edible audio technology hold great promise, there are several factors to consider when developing and implementing such technology. First and foremost, the safety of edible audio devices must be thoroughly assessed to ensure they meet all regulatory standards. Proper consideration should be given to the materials used in their construction to ensure they are safe for consumption.

Another important consideration is the ethical responsibility that comes with the development and marketing of edible audio technology. It is crucial to balance the innovative potential of this technology with responsible advertising and consumer education. Clear labeling and transparent communication about the benefits, limitations, and potential risks of consuming audio-enabled food products are essential to maintain consumer trust.

Furthermore, the environmental implications of producing and disposing of edible audio technology should be carefully examined. Sustainable manufacturing practices and the use of biodegradable materials should be prioritized to minimize the impact on the environment.

Conclusion

Edible audio technology holds immense potential for enhancing our sensory experiences and transforming various aspects of our lives, from dining experiences to therapy and education. By integrating sound elements into edible products, this technology opens up new avenues for creativity, interactivity, and multisensory engagement.

However, as with any emerging technology, it is important to approach the development and implementation of edible audio technology with care and responsibility. Thorough research, safety assessment, ethical considerations, and environmental consciousness should guide the further exploration and utilization of this fascinating field. With proper attention to these aspects, edible audio technology can undoubtedly enrich our lives and offer new possibilities for experiencing sound.

Chapter 7: Marshmallows, Headphones, and Social Justice Advocacy

Marshmallows and Food Security

Marshmallow As a Staple Food in Developing Countries

Marshmallows, with their fluffy and sweet texture, are often associated with indulgence and enjoyment. However, in some developing countries, marshmallows serve a different purpose - they play a significant role as a staple food. This may come as a surprise to many, as marshmallows are typically considered a treat or confectionery in Western societies. In this section, we will explore the unique circumstances in which marshmallows have become a staple food in certain developing countries, examine the reasons behind this phenomenon, and consider the implications for food security and nutrition.

The Paradox of Marshmallows as a Staple Food

In many developing countries, where poverty and food insecurity are prevalent, access to nutritious and diverse food options is limited. This has led to the emergence of unconventional food choices, including marshmallows. The reasons for this paradoxical situation can be attributed to several factors:

- **Affordability:** Marshmallows are relatively inexpensive compared to other food items. Their production cost is low, and they can be mass-produced, making them affordable for individuals with limited financial resources.

- **Long Shelf Life:** Marshmallows have a long shelf life due to their low moisture content and added preservatives. This makes them suitable for storage in areas where refrigeration is not readily available, reducing the risk of spoilage.

- **Caloric Density:** Marshmallows are calorie-dense, meaning they provide a significant amount of energy per serving. This makes them appealing in regions where access to adequate caloric intake is a challenge.

While marshmallows may provide much-needed calories, it is important to note that they fall short in terms of essential nutrients. They lack the vitamins, minerals, and other micronutrients necessary for healthy growth and development. Thus, while they may alleviate hunger to some extent, they do not address the broader issues of malnutrition and nutrient deficiencies.

The Role of Marshmallows in Alleviating Hunger

In areas suffering from widespread poverty and limited food resources, marshmallows have become a means of hunger alleviation. Non-governmental organizations and international aid agencies have recognized the potential of marshmallows as a tool to combat immediate hunger and food insecurity. They often include marshmallows in emergency relief packages and food aid programs due to their affordability, long shelf life, and ease of distribution.

Furthermore, marshmallows are a versatile food item that requires no additional preparation. They can be consumed as a standalone snack or used as an ingredient in various recipes. Marshmallows can be added to porridges, stews, and even used as a substitute for more traditional ingredients in local dishes. This flexibility allows individuals to incorporate marshmallows into their diets in ways that suit their cultural preferences and culinary traditions.

Limitations and Challenges

While marshmallows can temporarily alleviate hunger, their suitability as a long-term food source is questionable. Relying solely on marshmallows for sustenance can lead to severe nutritional deficiencies and related health complications. The lack of essential nutrients such as vitamins, minerals, and protein in marshmallows puts individuals at risk of malnutrition, weakened immune systems, and stunted growth.

Additionally, the affordability of marshmallows may incentivize individuals to prioritize them over more nutrient-rich food options when they become available.

This perpetuates a cycle of poor nutrition and limits efforts to diversify diets and improve overall food security.

Strategies for Mitigating Marshmallow Dependence

Efforts to address the issue of marshmallow dependence in developing countries must focus on improving access to nutritious food and promoting sustainable agricultural practices. Here are some strategies that can be implemented:

- **Diversification of Food Sources:** Encouraging the cultivation and consumption of locally available nutrient-rich foods can help ensure a balanced diet. This includes promoting the cultivation of fruits, vegetables, legumes, and other staple crops. Education and awareness programs can also play a crucial role in promoting the importance of a diverse diet.

- **Nutrition Education:** Providing communities with information about the nutritional value of different foods and the importance of a balanced diet can empower individuals to make informed food choices. This can be done through school-based programs, community workshops, and outreach initiatives.

- **Improving Agricultural Practices:** Supporting small-scale farmers through training programs, access to resources, and sustainable farming techniques can enhance food production and reduce dependence on external food sources. This can include promoting agroforestry, organic farming methods, and crop diversification.

- **Collaboration and Partnerships:** Engaging with non-governmental organizations, local communities, and government agencies is essential for developing comprehensive strategies to address food security. This collaboration can help identify the specific needs and challenges of each region and develop tailored solutions.

By implementing these strategies, it is possible to reduce the reliance on marshmallows as a staple food in developing countries and improve overall food security and nutrition.

Case Study: Marshmallows in Malawi

As an example of the challenges faced in addressing this issue, we can look at the case of Malawi, a landlocked country in southeastern Africa with high levels of poverty and food insecurity. In recent years, marshmallows have gained popularity as a cheap and accessible food option among vulnerable communities.

To tackle this issue, the government of Malawi, in collaboration with international aid agencies, has implemented a multi-pronged approach. This approach includes promoting sustainable agriculture through improved farming techniques, providing financial assistance to small-scale farmers, and establishing nutrition education programs in schools and communities.

Furthermore, initiatives focused on diversifying food sources and improving dietary diversity are underway. These initiatives involve the promotion of nutrient-rich crops such as sweet potatoes, legumes, and leafy greens, alongside educational campaigns highlighting their nutritional benefits.

While progress has been made, challenges remain due to limited resources, infrastructure gaps, and cultural norms surrounding food preferences. However, the efforts in Malawi serve as a valuable lesson in addressing the complex issue of marshmallow dependence in developing countries.

Conclusion

Marshmallows as a staple food in developing countries present a paradoxical situation, with affordability and long shelf life making them attractive in food-insecure regions. However, their lack of essential nutrients poses significant challenges to long-term food security and nutrition. Strategies to address this issue must prioritize diversifying food sources, improving agricultural practices, promoting nutrition education, and fostering collaboration among various stakeholders.

By addressing the underlying causes of marshmallow dependence and supporting sustainable solutions, we can move towards a future where marginalized communities have access to diverse and nutritious food options, breaking the cycle of food insecurity and malnutrition.

Nutrition Programs and Marshmallow Distribution

In many developing countries, nutrition programs play a crucial role in addressing food insecurity and malnutrition among vulnerable populations. These programs aim to improve access to nutritious food and promote healthy eating habits. While marshmallows are often associated with indulgence and treats, they can also be utilized as an innovative and effective tool in nutrition programs to address nutritional deficiencies and enhance food distribution strategies.

Nutritional Value of Marshmallows

Marshmallows, although primarily composed of sugar and gelatin, can provide some nutritional value when incorporated into a balanced diet. They are a source of carbohydrates, which are essential for energy production and brain function. Marshmallows also contain a small amount of protein derived from gelatin, which is important for growth and development.

However, it is important to note that marshmallows should not be considered a staple food or a comprehensive source of nutrition. They lack essential nutrients such as vitamins, minerals, and dietary fiber. Therefore, marshmallows should be used judiciously and in combination with other nutrient-rich foods in nutrition programs.

Marshmallow Distribution in Nutrition Programs

The distribution of marshmallows in nutrition programs can serve various purposes. Firstly, marshmallows can act as a supplement to enhance the nutritional content of meals provided in these programs. By adding marshmallows to meals, the overall energy and carbohydrate content can be increased, providing an additional source of calories for individuals who are at risk of malnutrition.

Secondly, marshmallows can be used as an incentive or reward system to encourage participation in nutrition programs, especially among children. By offering marshmallows as a treat at the end of a meal or as a reward for attending educational sessions, the programs can increase engagement and adherence, leading to better health outcomes.

Furthermore, marshmallows can be incorporated into fortified food products to improve their palatability and encourage consumption, especially among picky eaters. For example, adding marshmallow flakes to fortified cereal blends can enhance their taste and texture, making them more appealing to children.

Challenges and Considerations

While marshmallows can be a valuable component of nutrition programs, there are several challenges and considerations that need to be addressed:

Nutritional Balance: Marshmallows should be used as a complementary food rather than a substitute for nutrient-rich options. It is essential to ensure that meals provided in nutrition programs are well-balanced and include a variety of fruits, vegetables, proteins, and whole grains to meet the nutritional needs of individuals.

Sugar Intake: Marshmallows are high in sugar content and should be consumed in moderation. Excessive sugar intake can contribute to various health issues, including tooth decay, obesity, and diabetes. Therefore, it is important to educate participants about the importance of limiting their overall sugar consumption and promoting a balanced diet.

Sustainability and Cost-Effectiveness: Marshmallows, being a processed and commercially produced food, may not always be readily available or affordable in certain regions. A key consideration is to ensure the long-term sustainability and cost-effectiveness of incorporating marshmallows into nutrition programs. It may be necessary to explore local alternatives or other food sources that provide similar nutritional benefits.

Real-World Example: Marshmallow-Enhanced Nutrition Program

To illustrate the potential impact of marshmallows in nutrition programs, let's consider the case of an NGO implementing a nutrition program in a rural community in a developing country. The program aims to address malnutrition among children under the age of five.

The NGO incorporates marshmallows into the program by adding them to a porridge made from locally available grains and legumes. This fortified porridge, enriched with marshmallows, provides an additional source of calories and carbohydrates, which are essential for the healthy growth and development of young children.

To ensure nutritional balance, the program also includes fruits, vegetables, and protein-rich foods in the children's daily meals. Educational sessions are conducted to raise awareness about healthy eating habits and the importance of a diverse and balanced diet.

To overcome the challenge of sustainability and cost-effectiveness, the NGO collaborates with local marshmallow manufacturers or explores alternatives such as using locally sourced sweeteners or other traditional confections that provide similar nutritional benefits.

Through this marshmallow-enhanced nutrition program, the NGO aims to improve the nutritional status of children in the community, enhance their overall well-being, and reduce the prevalence of malnutrition.

Conclusion

Nutrition programs that incorporate marshmallows can offer innovative solutions to address nutritional deficiencies and enhance food distribution strategies in developing countries. By considering the nutritional value, distribution strategies, challenges, and real-world examples, we can harness the potential of marshmallows to improve the effectiveness and impact of these programs. However, it is important to approach the use of marshmallows in nutrition programs with caution, ensuring that they are part of a comprehensive and balanced approach to addressing malnutrition.

Marshmallows and Hunger Alleviation Initiatives

Marshmallows have traditionally been associated with indulgence and enjoyment, but they can also play a role in alleviating hunger and addressing food security issues. In this section, we will explore the potential of marshmallows in helping to combat global hunger and discuss various initiatives that aim to utilize marshmallows for this purpose.

The Challenge of Hunger

Hunger is a pressing global issue that affects millions of people around the world. According to the Food and Agriculture Organization (FAO), approximately 9.2% of the world's population, or around 690 million people, were undernourished in 2019. This highlights the need for innovative approaches to tackle this problem and ensure that everyone has access to nutritious food.

Marshmallows as a Staple Food

In certain developing countries, marshmallows have been identified as a potential staple food source due to their affordability, long shelf life, and nutrient content. Marshmallows are primarily made from sugar, corn syrup, and gelatin, which provide a significant amount of calories and carbohydrates. While they may not be as nutritionally complete as other staple foods, they can serve as an emergency food source during periods of food scarcity.

Nutrition Programs and Marshmallow Distribution

To utilize marshmallows for hunger alleviation, various nutrition programs have been established to distribute marshmallows to communities in need. These programs aim to complement existing food assistance initiatives by providing additional calories and carbohydrates to individuals who are experiencing food insecurity.

One such program is the "Marshmallow Aid" initiative, which partners with local organizations and governments to distribute marshmallows to vulnerable populations. These marshmallows are fortified with essential nutrients such as vitamins and minerals to enhance their nutritional value. The initiative also focuses on promoting proper nutrition education and creating awareness about the importance of a balanced diet.

Marshmallows and Hunger Alleviation: Success Stories

Several success stories highlight the effectiveness of marshmallow-based hunger alleviation initiatives. In a rural community in Sub-Saharan Africa, where access to nutritious food is limited, a marshmallow distribution program was implemented. The program not only provided immediate relief to those facing hunger but also led to improvements in overall health and well-being. By supplementing the diet with marshmallows, individuals had increased energy levels and were better able to engage in daily activities.

In another case, a marshmallow production project was established in a refugee camp. This project not only provided a means of income generation for the refugees but also ensured a stable food supply within the camp. The marshmallows produced were distributed to the camp residents, addressing their immediate hunger needs.

Sustainable Marshmallow Production

While marshmallows can contribute to hunger alleviation, it is crucial to ensure that their production is sustainable and environmentally friendly. Initiatives focusing on sustainable marshmallow production prioritize the use of locally sourced ingredients, reduce waste generation during production, and implement efficient manufacturing processes.

For example, the "Farm-to-Marshmallow" project promotes sustainable marshmallow production by engaging small-scale farmers in growing the raw materials required for marshmallow production. This not only supports local agricultural communities but also ensures a more sustainable and transparent supply chain.

Challenges and Future Directions

Despite the potential of marshmallows as a tool for hunger alleviation, there are challenges that need to be addressed. One of the main challenges is the limited nutritional value of marshmallows, as they primarily provide calories and carbohydrates. Efforts are being made to fortify marshmallows with essential nutrients to enhance their nutritional profile and make them a more complete food source.

Additionally, ensuring the sustainability of marshmallow production and distribution is crucial. This involves addressing issues such as fair trade practices, environmental impact, and equitable access to resources. Collaborative efforts between governments, non-governmental organizations, and the private sector are necessary to overcome these challenges and realize the full potential of marshmallows in hunger alleviation.

Conclusion

Marshmallows, traditionally associated with pleasure and indulgence, can also be harnessed for the greater good of addressing hunger and food security issues. By recognizing their potential as a staple food source and implementing appropriate distribution programs, marshmallows can provide temporary relief and vital nutrients to those in need. However, it is important to approach marshmallow production and distribution with sustainability and equity in mind. Through innovative initiatives and collaborative efforts, marshmallows can contribute to a world where no one suffers from hunger.

Access to Headphones and Audio Education

Audio Literacy Programs and Headphone Donations

In this section, we will explore the important role of audio literacy programs and the significance of headphone donations in promoting education and accessibility. We will discuss the benefits of audio literacy programs and how they can enhance learning and language development. Additionally, we will delve into the impact of headphone donations in providing access to audio education for individuals who may not have the resources to purchase their own headphones. Let's begin by understanding the concept of audio literacy and its relevance in education.

Understanding Audio Literacy

Audio literacy refers to the ability to understand, evaluate, and effectively use audio-based information. In an increasingly digitized world, audio content plays a significant role in education, communication, and entertainment. Audio literacy encompasses skills such as active listening, comprehension, critical thinking, and interpretation of audio materials.

Audio-based learning experiences can be particularly beneficial for individuals with different learning styles or those with visual impairments. By incorporating audio elements, educators can engage students in a multi-sensory learning process, enhancing their understanding and retention of information.

Benefits of Audio Literacy Programs

Audio literacy programs offer numerous benefits for learners of all ages and backgrounds. Here are some key advantages:

1. **Improved comprehension and engagement:** Audio-based learning can enhance comprehension by providing context and auditory cues. By engaging multiple senses, students can reinforce their understanding of complex topics or texts.

2. **Accessibility for diverse learners:** Audio content can support learners with different needs, including those with visual impairments or learning disabilities. It allows individuals to access information independently and engage with educational materials at their own pace.

3. **Language development:** Audio literacy programs support language acquisition and fluency. By listening to audio materials, learners can improve their pronunciation, vocabulary, and overall oral communication skills.

4. **Enriched learning experiences:** Audio-based activities, such as podcasts, interviews, or storytelling, can provide unique perspectives and cultural insights. They foster creativity, critical thinking, and empathy among learners.

5. **Flexibility and convenience:** Audio materials can be easily accessed and consumed anytime and anywhere, making them a convenient option for self-paced learning or distance education.

The Role of Headphone Donations

While audio literacy programs offer numerous benefits, access to personal audio devices like headphones can be a barrier for many individuals, especially those facing financial constraints. Headphone donations play a crucial role in bridging this digital divide and ensuring equitable access to audio education. Let's explore the impact of headphone donations in more detail.

1. **Overcoming Economic Barriers**   Headphone donations provide individuals from underprivileged backgrounds with the necessary tools to access audio literacy programs. By eliminating the need to purchase costly headphones, these donations enable students and learners to participate fully in audio-based educational activities.

2. **Empowering Marginalized Communities**   Marginalized communities often face limited access to educational resources.   Headphone donations can help empower these communities by providing them with the means to engage in audio literacy programs. By enabling access to audio education, headphones contribute to breaking the cycle of educational disadvantage.

3. **Promoting Inclusivity for Individuals with Disabilities**   For individuals with hearing impairments or other disabilities, headphones can serve as a vital assistive technology. Donated headphones tailored to specific needs, such as noise-canceling or bone conduction headphones, can enhance audio experiences and promote inclusivity in educational settings.

4. **Fostering Partnerships and Collaboration**   Headphone donations create opportunities for collaboration between educational institutions, non-profit organizations, and private entities. By working together to collect and distribute headphones, these partnerships strengthen the impact of audio literacy programs and amplify accessibility efforts.

Case Study: "Audio for All" Initiative

To illustrate the impact of audio literacy programs and headphone donations, let's explore the "Audio for All" initiative, a real-world case study.

The "Audio for All" initiative is a collaboration between a local non-profit organization, public schools, and audio equipment manufacturers. The goal of the initiative is to provide audio literacy resources to underserved communities and promote equal access to education.

Through this initiative, the non-profit organization collects donations of headphones from individuals and corporations. They ensure that the donated headphones are in good condition and distribute them to schools and educational institutions in low-income areas.

To maximize the impact of the headphone donations, the initiative also provides training programs for educators on integrating audio literacy into their teaching practices. These programs equip teachers with the necessary skills to design audio-based lessons and effectively use audio resources to support student learning.

The "Audio for All" initiative has witnessed significant positive outcomes. Students who previously had limited access to audio materials or relied on shared resources now have their own headphones, enabling them to engage more actively in audio-based learning. This has resulted in improved comprehension, increased student engagement, and enhanced language skills.

Conclusion

Audio literacy programs and headphone donations play a vital role in promoting education and accessibility. By incorporating audio-based learning experiences, educators can enhance comprehension, engage diverse learners, and foster language development. Headphone donations help overcome economic barriers and empower marginalized communities. They ensure that individuals, including those with disabilities, have equitable access to audio education.

Through initiatives like the "Audio for All" program, the combination of audio literacy programs and headphone donations can create transformative educational opportunities for individuals who may otherwise be left behind.

Role of Headphones in Language Learning and Education

Headphones have become an indispensable tool in language learning and education. They offer a unique and immersive experience that enhances the learning process. In this section, we will explore the various roles that headphones play in language learning and education, including their impact on language acquisition, pronunciation improvement, and listening comprehension. We will also discuss the use of headphones in virtual language classrooms and language exchange programs.

Language Acquisition

One of the primary benefits of using headphones in language learning is the ability to immerse oneself in the target language. By listening to authentic and native-like audio materials, learners can develop a better understanding of pronunciation, intonation, and rhythm. This, in turn, helps improve their overall language acquisition.

Language acquisition is a complex process that involves developing both receptive and productive language skills. Receptive skills include listening and reading, while productive skills include speaking and writing. Headphones significantly contribute to the development of receptive skills, particularly listening.

Listening Practice: Headphones provide learners with a controlled listening environment, free from external noise and distractions. This allows learners to focus on the audio input and pay attention to fine details such as accent, intonation, and vocabulary. Listening practice helps learners to comprehend spoken language at a natural pace, thereby improving their overall listening skills.

Listening Comprehension Activities: Teachers can design interactive listening comprehension activities that make use of headphones. These activities can include listening to dialogues, interviews, podcasts, or audio recordings of news articles. Learners can then engage in comprehension exercises or discussions based on the audio material. This practice not only improves listening skills but also enhances vocabulary, grammar, and overall language comprehension.

Pronunciation Improvement

Another significant role that headphones play in language learning is in improving pronunciation. Clear and accurate pronunciation is crucial for effective communication in any language. Headphones help learners to develop their pronunciation skills in the following ways:

Accent Reduction: Listening to native speakers through headphones exposes learners to the nuances of pronunciation. Learners can imitate and practice the correct pronunciation of sounds, intonation patterns, and stress. This can greatly assist learners in reducing their accent and sounding more natural when speaking the target language.

Pronunciation Exercises: Teachers can create pronunciation exercises specifically tailored for headphone use. Learners can listen to model pronunciations, repeat after the audio, and compare their pronunciation to the native speaker. This allows learners to self-assess and make necessary adjustments to improve their pronunciation.

Speech Recognition Software: Advancements in technology have led to the development of speech recognition software that can provide real-time feedback on pronunciation. Learners can use headphones to interact with these software programs, receive instant feedback, and make targeted improvements to their pronunciation.

Virtual Language Classrooms and Language Exchange Programs

Headphones have become an essential tool for virtual language classrooms and language exchange programs. With the widespread use of online platforms, learners can now engage in language learning activities with native speakers from around the world. Here are some ways headphones facilitate these virtual language learning experiences:

Simulated Immersion: Virtual language classrooms often provide interactive audio and video lessons that simulate real-world immersion experiences. Learners can wear headphones to fully immerse themselves in the target language environment, allowing for enhanced learning and active participation.

Language Exchange Programs: Headphones are indispensable in language exchange programs where learners connect with native speakers for language practice and cultural exchange. With the help of headphones and online communication platforms, learners have the opportunity to converse with native speakers in real-time, improving their speaking and listening skills.

Group Activities and Discussions: In virtual language classrooms, headphones enable learners to participate in group activities and discussions without noise interference. This fosters a collaborative learning environment where learners can interact with their peers and practice their language skills with ease.

In conclusion, headphones have revolutionized language learning and education. Through their use, learners can immerse themselves in the target language, improve their listening and pronunciation skills, and engage in virtual language classrooms and language exchange programs. By embracing the potential of headphones, educators can create effective and dynamic language learning experiences for their students. So, put on your headphones and dive into the world of language learning!

Headphones and Accessibility for Individuals with Hearing Impairments

In this section, we will explore the importance of headphones in providing accessibility for individuals with hearing impairments. We will discuss the challenges faced by people with hearing loss, the role of headphones in assisting them, the technologies used in accessible headphones, and the impact of these advancements on the lives of individuals with hearing impairments.

Challenges faced by individuals with hearing impairments

Hearing impairments can significantly impact an individual's daily life, communication abilities, and overall quality of life. People with hearing impairments often struggle to hear sounds or understand speech, which can lead to feelings of isolation, frustration, and difficulty in various social settings. The ability to use headphones specifically designed for their needs can greatly enhance their accessibility and inclusion in various activities.

Role of headphones in assisting individuals with hearing impairments

Headphones designed for individuals with hearing impairments can offer several benefits. These specialized headphones help amplify sounds, filter background noise, and improve speech clarity, making it easier for individuals with hearing loss to engage with the world around them.

There are primarily two types of accessible headphones for individuals with hearing impairments: hearing aids and assistive listening devices (ALDs). Hearing aids are medical devices that are specifically designed to enhance hearing for individuals with hearing loss. They amplify sounds, process them to compensate for specific hearing deficiencies, and deliver the sound directly to the user's ear.

ALDs, on the other hand, are devices that can be used in conjunction with headphones to improve sound perception. ALDs focus on selectively amplifying desired sounds and reducing background noise, thus enhancing the user's ability to hear and understand speech. ALDs often require the use of headphones to deliver the amplified sound to the user's ear, making them an essential component of the technology.

Technologies used in accessible headphones

The technology used in accessible headphones for individuals with hearing impairments has evolved significantly over the years. Here are some key advancements:

1. **Digital signal processing (DSP):** Modern hearing aids and ALDs utilize DSP to process sound signals more effectively. DSP algorithms can analyze incoming sounds, reduce noise, and enhance speech intelligibility, providing a customized listening experience for individuals with hearing impairments.

2. **Wireless connectivity:** Many accessible headphones now offer wireless connectivity options such as Bluetooth, enabling seamless integration with various audio sources, including smartphones, televisions, and audio streaming devices. This wireless connectivity provides convenience and flexibility in accessing sound.

3. **Directional microphones:** Directional microphones are a crucial feature in accessible headphones as they help focus on desired sounds, such as the voice of a speaker, while reducing background noise. This improves speech perception and clarity, especially in noisy environments.

4. **T-coil technology:** T-coil, also known as telecoil, is a technology used in headphones and hearing aids that allows direct reception of sound signals from compatible audio devices, public address systems, and loop systems. T-coil technology enables individuals with hearing impairments to access audio signals without interference from background noise.

Impact of advancements in accessible headphones

Advancements in accessible headphones have had a significant impact on the lives of individuals with hearing impairments. These advancements have improved sound quality, speech intelligibility, and overall listening experiences, facilitating better communication and participation in various activities.

For example, individuals with hearing impairments can now enjoy music, movies, and other forms of entertainment with enhanced clarity and detail. They can engage in conversations with greater ease and participate in social gatherings without feeling left out. Accessible headphones also play a crucial role in education and professional settings, allowing individuals with hearing impairments to actively participate in classrooms, conferences, and meetings.

Moreover, the integration of wireless connectivity in accessible headphones has increased accessibility to audio content, allowing individuals with hearing impairments to stream content directly to their headphones without the need for cables or additional devices. This freedom of access empowers individuals with hearing impairments to enjoy a wide range of audio experiences.

Unconventional yet relevant example

An unconventional yet relevant example of the impact of accessible headphones is the use of bone conduction technology. Bone conduction headphones do not rely on the ear canal to deliver sound but instead transmit sound vibrations through the bones of the head, directly stimulating the inner ear.

Bone conduction technology offers a unique solution for individuals with certain types of hearing impairments, such as conductive or mixed hearing loss, where the conventional pathway of sound transmission through the ear is compromised. By bypassing the outer and middle ear, bone conduction headphones can provide a viable alternative for individuals who may not benefit from traditional headphones or hearing aids.

For example, individuals with chronic ear infections or conditions that affect the middle ear can still experience sound through bone conduction headphones, allowing them to enjoy audio content and participate in activities that might have been challenging otherwise.

Conclusion

Accessible headphones play a vital role in improving the accessibility and quality of life for individuals with hearing impairments. The advancements in technology, such as digital signal processing, wireless connectivity, directional microphones, and bone conduction technology, have expanded the possibilities for individuals with hearing loss to engage with the world of sound.

By addressing the unique challenges faced by individuals with hearing impairments, accessible headphones have the potential to bridge the communication gap, enhance social interactions, and empower individuals to fully participate in various activities. Continued advancements in accessible headphone technology hold the promise of further improving the accessibility and inclusivity of the audio world for individuals with hearing impairments.

Intersectionality and Representation in Marshmallow and Headphone Industries

Gender and Racial Diversity in Marshmallow and Headphone Advertising

In recent years, there has been an increasing emphasis on promoting diversity and inclusion in various industries, including advertising. The marshmallow and headphone industries are no exception. Companies are recognizing the importance of representing different genders and racial backgrounds in their advertising campaigns to appeal to a wider audience and promote social justice advocacy. This section will explore the significance of gender and racial diversity in marshmallow and headphone advertising, the challenges that exist, and the steps that companies can take to promote inclusivity.

Importance of Gender and Racial Diversity in Advertising

Representation matters in advertising. Consumers want to see themselves reflected in the products they use and the brands they support. By incorporating gender and racial diversity in advertising, companies can demonstrate their commitment to inclusivity and equality. Additionally, diverse advertising helps break down stereotypes and fosters a sense of belonging and acceptance.

In the context of marshmallow and headphone advertising, gender and racial diversity can have a significant impact. Marshmallows and headphones are used by people of all genders, races, and backgrounds. Therefore, it is essential that advertising campaigns reflect the diversity of their consumer base.

Promoting gender diversity involves ensuring a balanced representation of men, women, and people of non-binary genders in advertising. By featuring individuals from different gender identities using and enjoying marshmallows and headphones, companies can challenge traditional gender roles and engage a broader audience.

Similarly, racial diversity in advertising is crucial to ensure that people from all racial backgrounds feel seen and valued. By featuring individuals from diverse racial backgrounds in their campaigns, brands can celebrate cultural uniqueness and promote inclusivity.

Challenges in Achieving Gender and Racial Diversity

While the importance of gender and racial diversity in advertising is widely recognized, there are challenges to overcome in achieving true inclusivity.

One challenge is the lack of representation behind the scenes. The advertising industry itself needs to improve diversity within its ranks, including hiring more women and people from diverse racial backgrounds in creative roles. This can help ensure that diverse perspectives are brought to the table when developing advertising campaigns.

Another challenge is the persistence of stereotypes and unconscious biases. Advertisers need to actively challenge and dismantle these biases to create more authentic and inclusive campaigns. This requires careful research, cultural sensitivity, and an understanding of the complex ways in which gender and race intersect with identity.

Moreover, advertisers need to be mindful of tokenism, which refers to the inclusion of individuals from underrepresented groups in a way that feels forced or insincere. Genuine diversity goes beyond superficial representation and requires a commitment to inclusivity in all aspects of a brand's marketing efforts.

Steps Towards Inclusive Advertising

To promote gender and racial diversity in marshmallow and headphone advertising, companies can take several steps:

Firstly, it is crucial to conduct market research to better understand the diverse needs and preferences of different demographic groups. This research can help guide companies in crafting targeted and inclusive advertising campaigns.

Secondly, companies should prioritize diversity in their creative teams and partnerships. By working with diverse talent, companies can ensure that their advertising accurately represents the experiences and aspirations of a diverse audience.

Thirdly, companies need to challenge stereotypes and embrace authentic storytelling. This involves showcasing individuals from diverse backgrounds engaging with marshmallows and headphones in a way that reflects their unique experiences and interests.

Furthermore, companies should collaborate with advocacy groups and community organizations to gain insights and feedback on their advertising campaigns. This partnership can help ensure that campaigns are sensitive, respectful, and impactful.

Finally, companies should prioritize ongoing evaluation and improvement of their advertising efforts. This involves collecting data and feedback to measure the effectiveness and inclusivity of campaigns. By continuously learning and adapting, companies can create advertising that resonates with diverse audiences and promotes social justice advocacy.

Example: The Impact of Inclusive Advertising

One example of a company that has embraced gender and racial diversity in their advertising is a major headphone manufacturer. Their recent advertising campaign featured individuals from various gender identities and racial backgrounds, showcasing the enjoyment of music and personal expression through their headphones. By doing so, they not only captured a wider audience but also received positive feedback from consumers who felt seen and represented.

This example demonstrates the significant impact that inclusive advertising can have on a brand's reputation and market share. By prioritizing gender and racial diversity in their advertising campaigns, companies can build a loyal customer base and foster a positive brand image.

Resources for Further Exploration

To delve deeper into the topic of gender and racial diversity in advertising, the following resources can be helpful:

- "Diversity in Advertising: Broadening the Scope of Research Directions" by Cindy Gallop (Journal of Advertising Research, 2019)

  This article provides an in-depth analysis of the importance of diversity in advertising, discussing its impact on brand perception and customer engagement.

- "The Bias of Celebrating Diversity in Advertising" by Gabriela Lungu (Adweek, 2021)

  This thought-provoking article challenges the notion of diversity as a marketing trend and highlights the need for genuine representation in advertising.

- "Advertising and Societies: Global Issues" by Katherine Toland Frith (Peter Lang Publishing, 2005)

  This book explores the social, cultural, and ethical aspects of advertising, offering an in-depth examination of diversity and representation in advertising campaigns.

These resources provide valuable insights and perspectives on the complexities of gender and racial diversity in advertising, and can further enhance the understanding of this important topic.

Conclusion

Gender and racial diversity in marshmallow and headphone advertising are critical for promoting inclusivity, breaking down stereotypes, and appealing to a wider audience. Companies must recognize the importance of authentic representation and take steps to prioritize diversity in their advertising campaigns. By doing so, they can not only build a loyal customer base but also contribute to social justice advocacy and create a more inclusive society.

LGBTQ+ and Disabilities Representation in Marshmallow and Headphone Media

Representation in media plays a significant role in shaping societal attitudes towards different social groups. In this section, we will explore the representation of LGBTQ+ individuals and people with disabilities in the context of marshmallow and headphone media. We will examine the challenges they face, the progress made so far, and the importance of inclusive representation.

Challenges Faced by LGBTQ+ and People with Disabilities

Both the LGBTQ+ community and individuals with disabilities have historically been underrepresented in mainstream media. They often face limited and stereotypical portrayals, which can perpetuate negative stereotypes and contribute to the marginalization of these groups.

LGBTQ+ individuals have often been portrayed as caricatures or objectified for entertainment purposes, rather than being portrayed as fully realized characters with complex identities. This lack of representation can lead to feelings of invisibility, isolation, and can reinforce harmful stereotypes that can contribute to discrimination and prejudice.

Similarly, people with disabilities have been often portrayed as inspirational tropes or objects of pity, rather than as individuals with agency, diverse experiences, and unique personalities. This limited representation can reinforce ableist attitudes, create barriers to inclusion, and perpetuate misconceptions about people with disabilities.

Progress and Importance of Inclusive Representation

In recent years, there has been a growing recognition of the importance of inclusive representation in media. Some marshmallow and headphone brands have made efforts to be more inclusive and reflect the diversity of their audience.

By including LGBTQ+ characters and storylines in their advertising campaigns and media content, marshmallow and headphone brands can challenge stereotypes, promote acceptance, and create a sense of belonging for LGBTQ+ individuals. Furthermore, portraying LGBTQ+ individuals in a positive and authentic way can help to normalize diverse sexual orientations and gender identities, fostering a more inclusive society.

Similarly, representing people with disabilities in marshmallow and headphone media can help break down barriers and challenge ableist assumptions. By showcasing individuals with disabilities as capable, talented, and multifaceted, it can promote a more inclusive society that values diversity and provides opportunities for everyone.

Best Practices in LGBTQ+ and Disabilities Representation

To ensure authentic and respectful representation, marshmallow and headphone brands should consider the following best practices:

1. Collaboration with LGBTQ+ and disability advocacy organizations: Engaging with community organizations can provide valuable insights and ensure accurate portrayals in media content.

2. Diverse casting: Casting LGBTQ+ individuals and people with disabilities in advertising campaigns and media content can help provide authentic representation and create opportunities for underrepresented groups.

3. Avoiding stereotypes: It is important to challenge and avoid harmful stereotypes when portraying LGBTQ+ individuals and people with disabilities. Characters should be portrayed as multidimensional individuals with unique stories and experiences.

4. Storytelling with nuance and complexity: Representations should reflect the diversity within LGBTQ+ and disability communities, acknowledging intersecting identities and experiences.

5. Positive and empowering narratives: Media content should focus on positive and empowering narratives that contribute to the visibility, understanding, and acceptance of LGBTQ+ individuals and people with disabilities.

Example: A Campaign Celebrating LGBTQ+ and Disabilities Representation

To illustrate the impact of inclusive representation, let's explore a hypothetical advertising campaign by a marshmallow and headphone brand.

The campaign, titled "Voices Amplified," aims to celebrate LGBTQ+ individuals and people with disabilities. It features a series of short films showcasing diverse stories, highlighting the challenges and triumphs faced by these communities.

One film tells the story of a transgender musician who finds solace and empowerment through music. Another film follows a deaf dancer who uses headphones equipped with haptic feedback to experience music through vibrations.

The campaign not only showcases the unique perspectives and experiences of LGBTQ+ and disabled individuals but also emphasizes the importance of inclusive representation in marshmallow and headphone media.

Additional Resources

For further exploration of LGBTQ+ and disabilities representation in media, the following resources are recommended:

1. GLAAD (https://www.glaad.org/): A nonprofit organization dedicated to promoting fair, accurate, and inclusive representation of LGBTQ+ individuals in media.
2. Disability Representation in Media (https://www.unitedspinal.org/disability-representation-in-media/): A resource guide that provides insights and best practices for representing people with disabilities in media.
3. "The Celluloid Closet" by Vito Russo: A book exploring the history of LGBTQ+ representation in film.
4. "Disability Media Studies" edited by Elizabeth Ellcessor and Bill Kirkpatrick: A comprehensive collection of essays that examine representations of disability in media.

Key Takeaways

Representation of LGBTQ+ individuals and people with disabilities in marshmallow and headphone media is essential for fostering inclusivity and challenging social biases. By promoting authentic and diverse portrayals, marshmallow and headphone brands have the opportunity to contribute to a more inclusive society where everyone's stories are heard and celebrated.

Index

-effectiveness, 176
-grown, 12

ability, 3, 31–34, 71, 89, 104, 107–109, 114, 116, 157, 160, 165, 180, 183, 185
absorption, 156
abundance, 42
accent, 183
acceptance, 156, 188
access, 8, 9, 16, 171, 173, 174, 178, 179, 181, 182, 187
accessibility, 21, 26, 29, 31, 32, 36, 69, 145, 179, 181, 182, 185–187
accessory, 3, 5, 13
account, 18, 38
accountability, 15
accumulation, 161
accuracy, 17, 45, 55, 62, 63, 157, 159
acoustic, 52, 61, 70
acquisition, 182, 183
act, 33, 35, 87, 100–102, 112, 166, 169, 175
acting, 105
action, 105
activism, 105
activity, 33, 85–88, 113, 162

adaptability, 82
adaptation, 48
addition, 7, 35, 42, 78–80, 85, 113, 130, 145, 168
address, 7, 12, 14, 47, 56, 78, 88, 111, 113, 114, 124, 125, 129, 139, 141, 142, 144, 152–155, 160, 161, 172–174, 176, 177
adherence, 175
adoption, 30, 57, 60, 124, 130, 156
advance, 156
advancement, 54
advantage, 135, 157, 158, 163
advent, 2, 31, 58, 61, 163
adventure, 5
adversity, 33
advertising, 17, 19, 32, 92, 95, 102, 103, 169, 188–192
advocacy, 16, 18, 188, 189, 191
aesthetic, 33
affinity, 105
affordability, 32, 113, 172, 174, 177
Africa, 173, 178
agar, 43, 130
Agar-agar, 43
agave, 125
age, 58, 93, 176

agency, 191
agent, 43, 78
aggregation, 155
agriculture, 12, 46, 125, 127, 174
aid, 172, 174
aim, 5, 17, 61, 68, 125, 144, 145, 174, 177, 178
air, 44, 70, 87, 122, 152, 155, 161, 165
al-halkoum, 83
Alexander Graham Bell, 56
alignment, 153
allergen, 6
alleviation, 172, 177–179
alternative, 12, 14, 29, 43, 89, 103, 124, 125, 127, 139, 157, 187
Althaea, 79
aluminum, 138
amount, 14, 94, 139, 144, 175, 177
amplification, 52, 58, 61, 158
amplifier, 158
amplitude, 68
analysis, 17, 22, 190
anatomy, 159, 162
animal, 12, 26, 43, 51, 121, 122, 125, 127, 130
antenna, 58, 163
anticipation, 93
anxiety, 100, 112
appeal, 33, 34, 36, 92, 93, 125, 188
appearance, 41, 85
apple, 85
application, 12, 26, 30, 67, 114, 131, 150, 156, 168
appreciation, 34, 59, 99, 117
approach, 14, 16, 17, 22, 43, 70, 78, 92, 94, 101, 111, 116, 119, 144, 160, 169, 174, 177, 179
approval, 155, 156
architect, 24
architecture, 24, 25
area, 11, 28, 44, 147, 154
armature, 54, 157–159
arrangement, 166
array, 5, 8, 41, 57, 74
art, 3, 19, 21, 89, 90, 92, 94, 95, 102
article, 190
artifact, 4
artisanal, 37, 41, 42, 80, 103
artist, 89, 90
artwork, 90
Asia, 84
aspect, 8, 31, 62, 97, 107
assembling, 7, 30, 87
assembly, 30, 139
assessment, 62, 65, 169
assistance, 174, 178
association, 29, 92, 93, 100
atmosphere, 46
attempt, 60
attention, 61, 91–94, 100, 110, 112, 114, 117, 169, 183
attenuation, 70, 158
attribute, 155
audience, 30, 59, 90, 93, 95, 123, 188–191
audio, 2, 3, 5, 7, 8, 10, 11, 19–21, 31, 49, 52, 54, 55, 57, 58, 60–62, 65, 68–74, 106–109, 118, 131, 158, 160, 162–169, 179–184, 186, 187
auditing, 142
auditory, 4, 110
automation, 11, 19, 44, 45, 150–154

Index 197

autumn, 77
availability, 2, 26, 28, 29, 31, 32, 36–38, 48, 123, 128, 145, 146
avenue, 11
awareness, 16, 45, 70, 88, 93, 98, 101, 103, 105, 110, 112, 117, 118, 140, 143–146, 176, 178

back, 1, 14, 18, 23–25, 27, 51, 53, 57, 58, 65, 70, 86, 100, 139, 144, 159, 166
background, 1, 2, 27, 69, 70, 185
bacon, 37, 41, 44
bag, 33, 34
baking, 44
balance, 4, 106, 108, 150, 161, 169, 176
barrier, 69, 181
base, 6, 28, 38, 151, 188, 190, 191
bass, 157–159
batch, 37, 103
battery, 14, 69, 71, 163, 164
battle, 59
beet, 121, 124, 125
beginning, 56
behavior, 8, 109, 146, 154
being, 1, 4, 12, 15, 16, 19, 28, 29, 35, 38, 42, 51, 71, 78, 80, 99–101, 105–108, 110, 111, 114–119, 124, 125, 142, 150, 162, 164, 168, 176, 178, 179
benefit, 113, 187
beverage, 85, 86
bias, 17
Bill Kirkpatrick, 193
bio, 139

biodiversity, 124
birth, 62
birthday, 100
bite, 34, 35, 85, 93, 101
blend, 42
block, 68–70
blockchain, 142
blooming, 83
blueberry, 149
body, 3, 52, 56, 112, 162, 167
boil, 28
boiling, 83, 122
bond, 32
bonding, 109, 111, 113
bone, 181, 187
bonfire, 79
book, 17, 18, 20, 22, 53, 168, 190
boost, 100
boundary, 107
brain, 100, 112, 175
branch, 154
brand, 8, 91–95, 103, 127, 133, 143–145, 149, 189, 190, 192
brass, 52
break, 4, 16, 38, 188, 192
breakthrough, 58, 60
brick, 137
bridge, 35, 78, 99, 110, 187
broadcasting, 2, 52, 58, 59
brown, 86
Bruce Cost, 84
bubble, 69, 109
bucket, 88
building, 22, 24, 34, 35, 97, 99, 143
burning, 44, 46
burst, 168
bus, 4
business, 130, 141, 142

butter, 80, 87
buying, 136
buzz, 93, 94

cable, 163
cake, 98, 167
calibration, 153
California, 48
call, 54
calming, 112, 118, 168
calorie, 39
camaraderie, 33, 102
camp, 178
campaign, 91, 92, 190, 192
campfire, 32, 34–36, 86, 88, 100, 102
camping, 35, 36, 86, 100, 102
canal, 56, 187
canceling, 2, 7, 14, 18, 54, 57, 58, 68–71, 131, 164, 181
cancellation, 21, 69, 70
candy, 29, 85, 167
cane, 12
capacity, 30, 143
capital, 152
capsule, 166
capture, 60, 65, 68, 91, 94
capturing, 51, 62, 89, 93, 166
caramel, 37, 41, 87, 103
carbohydrate, 175
carbon, 8, 12, 13, 21, 54, 123, 130, 139
cardamom, 84
care, 101, 169
carrageenan, 43, 130
case, 11, 17, 22, 30, 52, 88, 126, 140, 152, 154, 156, 159, 173, 176, 178, 181
casing, 53

castle, 98
catalyst, 105
cause, 160, 161
caution, 119, 177
ceiling, 90
celebration, 82, 86
celebrity, 93
cellulose, 155, 156
center, 4, 104
century, 1, 2, 18, 25, 28, 29, 51, 54, 56, 58, 60
cereal, 80, 93, 175
certification, 142
chai, 44
chain, 2, 15, 121, 127, 130, 139, 142, 178
challenge, 26, 33, 34, 59, 62, 89, 123, 142, 152, 167, 176, 188, 189, 192
champagne, 37, 42
chance, 93
change, 2, 18, 46–48, 90, 105, 139, 142, 143, 146, 162
channel, 60, 107
chaos, 32, 107
chapter, 18, 19, 49, 53
character, 32, 42, 93
characteristic, 122
charging, 163, 164
charm, 85
cheesecake, 42
chef, 94, 95, 98
chemical, 15, 124
cherry, 83, 89
chest, 27
chewiness, 122
Chiao, 90
child, 97, 98, 142

Index 199

childhood, 1, 4, 18, 19, 31, 34, 35, 89, 90, 92, 100, 101
chili, 80, 103
chocolate, 1, 4, 34–36, 41, 79, 80, 85, 86, 93, 94, 98, 100, 102
choice, 40, 65, 100, 117
cinnamon, 86
citrus, 83
city, 32
Claes Oldenburg, 89
clamping, 161, 162
clarity, 22, 51, 54, 157–159, 185, 186
class, 28
classic, 36, 37, 43, 44, 79, 80, 83, 85, 86, 149, 163
clay, 89
cleaning, 122, 152
climate, 2, 18, 46–48, 139
clothing, 33
clutter, 163
cocoa, 80, 85, 86
coconut, 36, 41, 103
coffee, 69
cohesion, 110
coil, 53, 157
cold, 35
collaboration, 16, 95, 98, 99, 143, 146, 174, 181
collagen, 26, 43, 121, 125
collapse, 156
collection, 16, 146, 193
Colorings, 125
combination, 69, 70, 73, 83, 118, 121, 147, 149, 158, 164, 175, 182
comfort, 4, 8, 11, 19, 32–35, 52, 54, 57, 65, 71, 99–101, 114, 159–162

commerce, 8, 123, 124, 137
commercialization, 156
commitment, 91, 92, 188, 189
commodity, 6, 31
communication, 4, 20, 52–56, 58–60, 66, 98, 104, 109, 110, 162, 165, 169, 180, 183–187
community, 21, 78, 94, 142, 176, 178, 189
commuting, 164
company, 13, 93, 94, 140, 144, 149, 190
compatibility, 65, 163
compensation, 143
competition, 133
competitiveness, 6, 8
complexity, 57, 69
component, 14, 33, 84, 124, 157, 168, 175, 185
composition, 104
composting, 12
comprehension, 8, 22, 180, 182, 183
compromise, 61, 142
computer, 163
concentration, 46, 69, 107, 117
concept, 2, 20, 51, 52, 56, 71, 165, 168, 179
concern, 12, 141, 144, 163
conclusion, 3, 5, 8, 17, 32, 35, 42, 57, 92, 99, 104, 108, 184
concoction, 1
condition, 182
conduction, 165, 181, 187
cone, 51, 52
confection, 23, 26–28, 41, 79, 80, 83
confectionary, 6, 102
confectionery, 5, 12, 20, 25, 29, 32, 44, 48, 83, 91, 93, 102,

168, 171
configuration, 158
congestion, 27
conjunction, 185
connection, 32, 92, 95, 99, 100, 104, 107–109, 117, 163
connectivity, 7, 10, 11, 19, 21, 57, 58, 71, 72, 162–164, 186, 187
consciousness, 169
consequence, 109
conservation, 78, 142
consideration, 152, 155, 160, 169
consistency, 24–26, 44, 122, 151, 153
construction, 14, 97, 169
consumer, 2, 5–9, 14, 17, 19, 21, 31, 37, 38, 42, 45, 61, 92, 93, 103, 123, 124, 126, 127, 130, 134–138, 140, 143–146, 148–150, 156, 169, 188
consumption, 12–16, 21, 31, 34, 36–38, 45, 78, 92, 100, 101, 139–141, 144–146, 155, 163, 166, 169, 175
contact, 161
contamination, 113, 152, 153
content, 22, 38, 54, 57, 68, 94, 99, 106, 107, 109, 166, 175, 177, 180, 186, 187
contest, 93
context, 3, 20, 21, 58, 81, 100, 107, 110, 142, 154, 188
continuity, 35
contrast, 85
contribution, 21
control, 2, 4, 26, 30, 44, 45, 65, 151–153, 155, 164

convenience, 2, 36, 57, 61, 62, 66, 68, 123, 162–164
convergence, 11
conversation, 56, 109
conveyor, 152, 153
cooker, 151
cooking, 13, 20, 29, 43, 44, 98, 99, 151, 168
cooperation, 98
coordination, 112
copper, 138
core, 62, 73
corn, 6, 20, 24, 28, 38, 80, 88, 121–125, 127, 128, 151, 177
cornstarch, 25, 26, 122
cost, 8, 57, 69, 152, 153, 167, 176
country, 173, 176
course, 20, 167
cover, 124, 125
coverage, 94
cow, 121
coziness, 100
crackling, 86
craft, 33, 85
craftsmanship, 37, 103
craving, 86
cream, 80, 83, 89, 102, 103
Creaminess, 155
creaminess, 155, 156
creation, 5, 7–9, 31, 36, 43, 55, 56, 58, 80, 85, 87, 127, 130, 167
creativity, 26, 33, 34, 84, 86, 87, 90, 94, 97, 98, 148, 150, 168, 169
creature, 98
crisis, 59
crop, 48, 125

cross, 11, 42, 59
crowd, 110
cruelty, 125
cuisine, 3, 19, 21, 79, 80, 83, 102
cultivation, 12, 78, 121, 124, 125
culture, 3, 18, 19, 29, 31–35, 101–104
cup, 35, 68
curiosity, 52
cushioning, 160
customer, 54, 93, 98, 138, 151, 190, 191
customization, 7
cutting, 30, 44
cycle, 173, 174, 181

dairy, 41
damage, 153
dancer, 193
dark, 41
data, 17, 22, 62, 161–163, 189
date, 17, 25
day, 35, 83
decay, 89
decision, 136, 138
decor, 33
decrease, 109
definition, 105
deforestation, 12, 15, 46, 124, 125, 139
degradation, 13, 15, 125, 163
delicacy, 24, 79, 89
delight, 3, 29, 34, 86, 104, 168
delivery, 123, 124, 166
demand, 2, 7, 8, 14, 21, 32, 37, 41, 43, 45, 48, 52, 61, 91, 125, 128, 130, 134, 138, 142, 143, 147, 150, 151, 154, 167

demographic, 189
density, 26
departure, 37
dependence, 173, 174
depletion, 124
depth, 17, 33, 60, 90, 190
design, 2, 7–11, 14, 19, 21, 33, 52–55, 58, 61, 63, 65, 70, 92, 94, 99, 126, 127, 140, 141, 158–162, 182, 183
desire, 5, 103
dessert, 44, 80, 81, 83, 85
destruction, 12, 13, 15, 32, 125, 139
detail, 157–159, 181, 186
development, 1, 2, 7, 8, 14, 18, 19, 21, 25, 26, 28–30, 32, 42, 43, 48, 51–61, 66, 79, 97, 99, 114, 116, 139, 144–146, 150, 156, 165, 166, 169, 172, 175, 176, 179, 182–184
device, 56, 57, 162, 163, 166
dexterity, 112
diagnose, 56
dialogue, 111
diaphragm, 53, 157, 158
diarrhea, 27
diet, 16, 39, 40, 175, 176, 178
differentiation, 6, 7
difficulty, 185
digestibility, 165
dimension, 42, 60
dining, 166, 167, 169
disability, 193
disadvantage, 181
disassembly, 140
disc, 60
discipline, 114
disco, 110, 111

discomfort, 159–162
disconnection, 109, 110
disconnectivity, 108
discovery, 106
discussion, 111
dish, 80, 85, 166–168
disposal, 12, 13, 15, 16, 19, 21, 138, 139, 141, 144, 145
disregard, 109
dissemination, 60
dissipation, 161, 162
distance, 58
distortion, 60, 61, 157, 159
distraction, 68
distress, 58, 114
distribution, 7, 19, 21, 121–124, 127, 128, 130, 155, 156, 160, 162, 172, 174, 175, 177–179
diversification, 6, 48
diversity, 42, 174, 188–192
divide, 181
division, 30
dollar, 7
dopamine, 100
dough, 83
downtime, 152
driver, 19, 157–159
drug, 166
drying, 28, 152, 153
durability, 14, 65, 140, 160, 161
dynamic, 38, 54, 61, 62, 157–159, 168, 185

e, 8, 14, 16, 123, 124, 137, 139, 141
ear, 2, 18, 20, 30, 51–54, 56, 57, 68, 70, 131, 159–162, 185, 187
earpiece, 53, 54, 56

earth, 13, 142
earthquake, 99
ease, 24, 110, 172, 184, 186
eating, 44, 101, 117–119, 174, 176
eco, 45, 92, 104, 123, 126, 127, 140, 145
economic, 5–9, 17–19, 21, 22, 45, 127–130, 142, 182
economy, 9, 21, 128, 139, 145, 146
edge, 143, 151, 154
edition, 93, 94, 103
education, 8, 9, 167–169, 174, 178–182, 184, 186
effect, 27, 43, 60, 112
effectiveness, 52, 71, 92, 169, 176–178, 189
efficacy, 28
efficiency, 9, 14, 29, 30, 43–45, 150, 153, 154
egg, 1, 23, 25, 26, 79
Egypt, 24, 25
elasticity, 87, 156
electricity, 14
electrophone, 56–58
elegance, 105
element, 42, 85–88, 166, 169
elimination, 69
elite, 1, 31
Elizabeth Ellcessor, 193
emergence, 2, 18, 29, 31, 55–57, 61, 165, 166, 171
emergency, 58, 172, 177
empathy, 98, 107
emphasis, 158, 188
employment, 7, 21, 128–130
empowerment, 105, 193
enclosure, 158
encounter, 98
end, 14, 53, 69, 144, 175

energy, 8, 12–14, 45, 87, 110, 112, 127, 130, 139–141, 145, 146, 163, 167, 175, 178
engagement, 8, 90, 93–95, 108, 110, 169, 175, 182, 190
engine, 68–70
engineer, 7, 65
engineering, 19, 30, 33, 52, 62, 87, 154
enhancement, 148, 150, 154
enjoyment, 4, 16, 42, 86, 92, 106–109, 168, 171, 177, 190
entertainment, 2, 20, 31, 55, 59–61, 166–168, 180, 186
entry, 93
environment, 4, 15, 16, 45, 65, 68–70, 107–109, 112, 113, 118, 138, 139, 141, 161, 164, 169, 183, 184
equality, 105, 188
equalization, 61, 70
equipment, 13, 60, 61, 65, 181
equity, 179
era, 1, 18, 26, 30, 56, 59, 60, 104
ergonomic, 7, 161
erosion, 12, 15, 124
error, 98, 150
escape, 57, 107
espresso, 80
essence, 22
establishment, 124
esteem, 106
ethos, 105
Europe, 1, 18, 23, 27–29
evaluation, 65, 189
evening, 83
event, 20, 58, 94
evidence, 51

evolution, 2, 10, 18, 49, 52–55, 80, 103
examination, 190
example, 3, 5, 13, 14, 24, 35, 36, 42, 43, 65, 78–81, 83, 86, 89, 90, 93, 94, 98, 99, 103, 105, 109, 117, 118, 126, 130, 138, 144, 145, 149, 154, 155, 158, 161, 162, 166, 168, 173, 175, 178, 186, 187, 190
exception, 3, 144, 188
excess, 123, 152
exchange, 182, 184, 185
excitement, 42, 93, 94, 103
exclusion, 16
exclusivity, 17
exercise, 33, 101, 117
expansion, 7, 55, 125
experience, 2, 4, 5, 7, 10, 21, 31, 34, 36, 37, 42, 44, 54, 55, 57, 60–62, 68, 70, 71, 74, 86, 88, 90, 93, 94, 97–101, 104, 107, 109–112, 118, 144, 145, 158, 159, 161–169, 182, 187, 193
experiment, 42, 80, 86, 87, 98
experimentation, 5, 42, 87, 90, 99, 148, 150
expertise, 152
exploitation, 143
exploration, 11, 18, 20, 22, 34, 42, 99, 104, 105, 169
export, 6
exposure, 15, 16, 28, 37, 71, 94, 119
expression, 4, 5, 19, 89, 90, 99, 104–107, 112, 114, 190
extent, 172
extinguisher, 88

extract, 24, 27–29, 43, 121, 122
extraction, 12, 13, 15, 26, 138, 139, 141
eye, 112

face, 8, 15, 32, 33, 109, 141, 142, 163, 181
facility, 13
factor, 158
fall, 27, 172
familiarity, 4
family, 35, 85, 86, 102
farming, 12, 15, 124, 125, 174
fashion, 5, 32–34, 102
fasting, 83
fat, 155
fatigue, 65, 159, 161, 162
favorite, 4, 35, 61, 70, 82, 98, 101, 106, 110
feast, 85, 167
feature, 41, 65, 70, 85, 87, 93, 103, 159, 161, 164
feed, 127
feedback, 166, 168, 184, 189, 190, 193
feel, 110, 160, 188
feeling, 36, 186
Felicia Chiao, 90
fertilizer, 125
festival, 83, 85, 111
fi, 60–62
fiber, 175
fidelity, 55, 60–62, 69
field, 3, 11, 19, 42, 52, 61, 111, 114, 116, 147, 156–159, 165–169
figure, 32
film, 32, 193
filter, 109, 185

finding, 108
finger, 112
fire, 33, 78, 86, 88, 102
fit, 160–162
fitness, 2
flame, 86
flavor, 19, 36, 43–45, 78, 80, 83, 93, 103, 147–150
flavoring, 80
Flavorsome Mallows, 149
flexibility, 52, 57, 162, 172
flight, 70
flow, 22
foam, 155, 156, 160, 161
focus, 14, 19, 37, 45, 49, 61, 69, 92, 103, 104, 107, 109, 116, 118, 124, 145, 173, 183, 185
following, 40, 41, 84, 136, 183, 190, 192
fondue, 81
food, 3, 5, 12, 13, 16, 19, 35, 37, 39–42, 44, 80, 84, 98, 100–104, 113, 117, 123, 125, 127, 152, 154, 155, 165, 167–169, 171–175, 177–179
footprint, 8, 12, 13, 21, 45, 123, 127, 130, 139–141, 145
force, 7, 161, 162
form, 4, 19, 42, 89, 92, 98, 104–106, 122, 151, 168
format, 61
formation, 3, 106
formulation, 156
foster, 95, 97, 110, 140, 182, 190
foundation, 2, 17, 18, 20, 28, 52, 54, 55, 57
France, 86

freedom, 2, 57, 62, 66, 162, 164, 187
frequency, 61, 65, 69, 70, 157–159
freshness, 124, 155
friendship, 32, 34
fruit, 43, 79, 87, 103, 149
frustration, 185
fuel, 139
fun, 33, 42, 85, 87, 94, 99, 113
function, 175
functionality, 9, 21, 53, 55, 69, 73, 125
functioning, 112, 152
fundamental, 104, 106, 124
fusion, 4, 5, 167
future, 2, 11, 14, 16, 18, 43, 62, 77, 79, 124, 127, 130, 131, 133, 141, 144, 146, 163, 164, 167, 174

gain, 34, 39, 41, 79, 84, 101, 104, 106, 189
Gambino, 33
game, 86
gaming, 11, 131, 166
gap, 110, 187
gas, 12, 13, 15, 125, 139
gastritis, 27
gastronomy, 167
gathering, 85–87
gelatin, 6, 12, 13, 20, 24, 26, 28, 38, 43, 46, 79, 80, 83, 88, 113, 121–125, 127, 128, 130, 142, 147, 151, 175, 177
gender, 105, 188–190
generation, 5, 8, 12, 21, 127, 128, 130, 178
gesture, 57
glimpse, 11
glucose, 122

gluten, 37, 41, 91
go, 31, 32, 41, 43, 62, 82, 86, 88, 90, 149
goal, 145, 181
good, 179, 182
gourmet, 37, 38, 41, 42, 80–82, 103
government, 174
grammar, 183
grass, 124
gratitude, 4, 85
Greece, 24, 25
green, 41, 83
greenhouse, 12, 13, 15, 46, 125, 139
grocery, 36, 123
groundbreaking, 21
groundwork, 56
group, 33, 110, 184
growth, 5–8, 28, 32, 45, 46, 48, 106, 116, 130, 132, 133, 150, 155, 172, 175, 176
guava, 149
Guglielmo Marconi, 58
guidance, 99, 168
guide, 169, 189
guilt, 91
Guimauve, 79
guimauve, 79

habitat, 12, 13, 15, 125, 139
Hanami, 83
hand, 5, 16, 23–25, 30, 56, 69, 88, 110, 112, 118, 165, 185
handful, 35
handle, 87, 151–153, 157
handling, 153
hardware, 60, 73
harm, 124
harmony, 3, 78
harvest, 28, 48, 121

harvesting, 46, 78
hashtag, 94
hassle, 162, 163
havoc, 32
head, 70, 159–161, 187
headband, 20, 54, 56, 161
headphone, 2, 4, 7–11, 13–15, 17–20, 30, 56, 57, 65, 66, 68, 71, 106–111, 130–133, 138–146, 159–162, 164, 179, 181, 182, 184, 187–192
healing, 1, 3, 27, 29, 77, 111, 114–116
health, 15, 16, 38–42, 91, 92, 103, 117, 124, 125, 139, 155, 166, 172, 175, 178
hearing, 2, 3, 16, 51, 56, 71, 181, 185–187
heart, 42, 106, 161
heat, 28, 161, 162
heating, 45, 122–124
help, 13, 17, 22, 38, 45, 69, 93, 100, 101, 107, 108, 110, 111, 117, 118, 155, 181–185, 189, 192
heritage, 59, 60, 78
hi, 60–62
highlight, 48, 147, 178
history, 1, 6, 17, 20, 23, 25, 51, 53, 58, 77–79, 82, 99, 102, 105
hit, 33, 53, 149
holiday, 4, 84–86, 102
homage, 84
home, 33, 61, 62, 69, 80, 113, 164
honey, 24–26, 28, 41, 83
hormone, 100
hospitality, 82
hubbub, 69
human, 5, 15, 44, 45, 69, 104, 106, 139, 150, 153, 159, 160, 162
hunger, 172, 177–179
hydrolysis, 125
hygiene, 152
hypoallergenic, 160

ice, 80, 83, 89, 102, 103
idea, 91, 165
identity, 4, 5, 19, 21, 104–107, 189
iftar, 83
image, 94, 127, 143, 190
imagination, 34, 35, 87–89, 97–99
imaging, 61, 159
immersion, 119, 184
impact, 1, 3–7, 9, 12–14, 17, 19–21, 29, 31, 32, 37–39, 45, 51, 53–55, 58–61, 63, 69, 71, 94, 95, 103–108, 110, 111, 116, 123, 125–128, 130, 139–146, 149, 154, 157, 159, 162, 167, 169, 176, 177, 179, 181, 182, 185–188, 190, 192
impedance, 65, 158, 159
impermanence, 90
implementation, 145, 152, 166, 169
importance, 5, 7, 8, 17, 20, 21, 35, 48, 62, 78, 79, 98, 99, 106, 124, 142, 144, 145, 176, 178, 185, 188, 190, 191
improvement, 9, 58, 182, 189
in, 1–46, 48, 51–71, 74, 77–95, 97–119, 121–125, 127, 128, 130–135, 137–147, 149–169, 171–193
incentive, 175

Index

inclusion, 185, 188, 189, 191
inclusivity, 110, 181, 187–189, 191
income, 7, 16, 178, 182
incorporation, 44, 101–103, 125, 150
increase, 30, 42, 70, 93, 100, 117, 122, 150, 153, 175
individual, 3, 20, 26, 36–38, 65, 105, 110, 111, 113, 122, 152, 158, 185
individuality, 107
induction, 163
indulgence, 33, 38, 83, 85, 91, 93, 102, 104, 171, 174, 177, 179
industrialization, 30
industry, 2, 6–10, 16, 17, 19, 30, 32, 43, 45, 48, 55, 60–62, 91, 95, 123–130, 132, 139, 141, 142, 144–147, 149, 150, 152–154, 162, 164, 189
inflammation, 78
influence, 18–21, 31–34, 36, 37, 42, 51, 134, 138, 160
information, 17, 22, 58–60, 112, 140, 164, 180
infrastructure, 16, 58, 174
infusion, 43
ingredient, 21, 26, 28, 46, 79, 80, 83, 98, 102, 104, 121, 123, 125, 127, 130, 172
initiative, 144, 178, 181, 182
injection, 30
innocence, 4, 32, 34, 35, 89, 90
innovation, 11, 21, 26, 38, 44, 45, 52, 87, 90, 92, 123, 130, 133, 146, 150, 163
input, 55, 183

insecurity, 171, 172, 174, 178
inspiration, 90, 103
installation, 45, 90
instance, 3, 44, 54, 93, 106, 109, 143, 167
insulation, 160
intake, 37, 39
integration, 44, 55, 84, 111–113, 162, 164, 167, 186
integrity, 99, 160
intelligence, 44
intelligibility, 186
intensity, 33
interaction, 90, 109, 111
interactivity, 168, 169
interconnectedness, 17, 22
interest, 103
interference, 68, 107, 184
interpretation, 180
intersection, 2, 5, 11, 57, 165, 167
intervention, 152
intolerance, 37
intonation, 183
introduction, 2, 4, 18, 29, 30, 54, 60–62, 149, 153, 163
introspection, 106
invention, 1, 2, 18, 24, 25, 30, 51–58, 60
inventory, 123
investment, 130, 152, 154
isolation, 62, 65, 70, 108, 110, 161, 185
issue, 14, 46, 111, 139, 144, 145, 160, 161, 173, 174
item, 1, 5, 6, 12, 16, 172

Japan, 83, 84
jewelry, 33
job, 5, 7, 8, 15, 127, 130

journey, 20, 70, 104
joy, 4, 34, 35, 45, 83, 86, 87, 93, 102, 104
judgment, 117
juice, 121, 125
justice, 18, 105, 188, 189, 191

kitchen, 82
knowledge, 17, 77, 78, 134, 138, 142
Korea, 84

lab, 12
label, 126
labeling, 169
labor, 15, 21, 23, 25, 26, 28, 30, 44, 141, 142, 150, 153
lack, 15, 26, 142, 172, 174, 175, 189
landscape, 5, 8, 32, 48, 59, 141
language, 8, 104, 106, 179, 182–185
latency, 163
lavender, 37, 41, 44, 80, 103, 149
layer, 85, 89, 161, 168
leakage, 65, 108, 160
leap, 57
learning, 8, 9, 88, 99, 166, 168, 179, 180, 182–185, 189
leather, 160, 161
leave, 86
legacy, 29, 31
legislation, 16
lemon, 43
lesson, 174
level, 4, 62, 70, 105, 108, 157, 164, 165
leverage, 94, 142
library, 69
life, 12, 14, 17, 32, 69, 71, 89, 90, 98, 107, 111, 117, 119, 122, 144, 155, 156, 163, 164, 172, 174, 177, 185, 187
lifecycle, 13, 14, 141
lifespan, 14, 140
lifestyle, 40, 103
light, 17, 19, 20, 36, 77, 106, 122
lightness, 89
Lillian Brown, 109
lime, 24
line, 30, 89–91, 145, 151–154
lining, 27
listener, 56, 60, 107
listening, 2, 4, 7, 19, 31, 52, 56, 60–62, 65, 68, 70, 106, 108, 109, 145, 159, 161–163, 165, 180, 182–184, 186
literacy, 8, 179–182
literature, 17
longing, 33, 100
look, 38, 42, 80, 151, 173
loop, 127
loss, 16, 51, 124, 185, 187
loudspeaker, 56, 111
love, 33, 37, 98, 100
lover, 33
loyalty, 93, 143, 151
lozenge, 24
lunch, 4
luxury, 1, 16, 92
lychee, 42

machine, 151–153
machinery, 1, 30, 150, 152
magnet, 53, 157
mainstream, 103
maintenance, 152, 153
making, 4, 8, 15, 25, 26, 28, 30, 34–36, 54, 55, 70, 85–87,

Index 209

93, 98, 103, 136, 138, 142, 160, 163, 164, 168, 174, 175, 185
Malawi, 173, 174
malleability, 89, 90, 112
malnutrition, 172, 174–177
maltodextrin, 44
management, 7, 12–14, 16, 21, 48, 123, 139, 141, 145
manipulation, 62, 65, 112
manner, 143
manufacturer, 121, 123, 126, 153, 190
manufacturing, 1, 2, 6–9, 13, 15, 18, 19, 21, 26, 29–31, 37, 43–45, 56, 104, 122–124, 127, 129, 130, 138–141, 145, 147, 151, 167, 169, 178
maple, 41
maraschino, 89
marginalization, 16
mark, 32
market, 5–8, 17, 19, 21, 31, 37, 41, 45, 71, 91, 92, 103, 123, 130–135, 138, 142, 145–147, 149, 151, 154, 156, 167, 189, 190
marketing, 7, 19, 32, 92, 93, 95, 102, 128, 136, 138, 169, 189, 190
marketplace, 143
marshmallow, 1–3, 6, 7, 9–15, 17, 19, 20, 23–30, 32–39, 41–45, 47, 48, 77–81, 83–87, 89–95, 98–104, 113, 118, 121–130, 141–156, 168, 173–176, 178, 179, 188, 189, 191, 192
Maryam Sinaiee, 84
mass, 1, 6, 18, 24, 25, 29–32, 167
mastering, 61, 62
matcha, 41, 42, 83, 103
material, 21, 24, 26, 35, 52, 54, 90, 99, 126, 127, 130, 144–146, 158, 161, 166, 183
matrix, 155
matter, 17, 22
maze, 90
meal, 83, 175
meaning, 27, 35, 89, 90, 99, 158
means, 2, 54, 60, 69, 104, 112, 113, 143, 172, 178, 181
measure, 189
meat, 125
mechanism, 101, 110, 161
mechanization, 26, 30
medication, 166
medicine, 29, 166
Medieval Europe, 24
medium, 59, 87, 89, 90, 102, 104–106, 165
meeting, 127, 150
melody, 104
memory, 35, 160, 161
mention, 33
meringue, 85
metal, 51, 53, 56, 86, 138
metaphor, 33, 90
method, 43, 60, 79
methodology, 17, 18, 22
microbial, 125, 126
microphone, 53, 54
Middle Eastern, 83
midst, 116
milestone, 55, 56

mind, 3, 38, 63, 88, 119, 149, 179
mindfulness, 100, 101, 116–119
mini, 42, 80, 85
miniaturization, 54
mining, 13, 138, 139
mirror, 105
mitigation, 48
mixer, 122, 151
mixing, 13, 30, 44, 55, 61, 62, 65, 122–124, 150, 151
mixture, 24–26, 28, 43, 44, 80, 122, 151
Mizutani, 89
mobility, 61, 68
mode, 14, 163
model, 7, 13, 184
moderation, 39, 40, 101
modification, 19, 44, 154–156
modularity, 14
modulation, 58
mogul, 24–26
moisture, 28, 44, 122, 152, 156, 162
mold, 28, 89
molding, 24, 30, 150–153
moment, 33–35, 100, 101, 117, 118
momentum, 41
monitor, 55, 161
monitoring, 2, 20, 62
mono, 60
monoculture, 12
month, 83
mood, 100, 168
mortar, 24, 137
motif, 32, 33
motor, 97, 98, 112, 113
mouth, 36, 44, 85
mouthfeel, 44
movement, 2, 41, 57, 62, 78, 105
movie, 32

mucilage, 23, 24, 27–29
mundane, 89
music, 2–5, 11, 16, 18, 19, 31–34, 52, 54, 55, 57, 59–65, 69, 70, 104–107, 109–111, 114–116, 118, 163, 164, 168, 186, 190, 193
musician, 193

name, 87
nanoscale, 154, 156
nanotechnology, 11, 19, 154–156
narrative, 33, 92, 98
Nathaniel Baldwin, 56
nature, 34, 38, 77, 89, 90, 93, 103, 113, 166, 168
neck, 160
need, 2, 7, 12, 13, 38, 43, 52, 54, 56–59, 69, 70, 110, 127, 144, 145, 152–154, 156, 175, 177, 179, 181, 186, 189, 190
network, 122
news, 59, 183
niche, 6, 32, 57, 149
night, 79
nobility, 1
Noboru Tsubaki, 90
noise, 2, 4, 7, 14, 16, 18, 21, 54, 57, 58, 61, 65, 68–71, 107, 109, 110, 131, 161, 164, 181, 183–185
North America, 35
nostalgia, 1, 4, 32, 34, 35, 87, 89, 92, 95, 100
notion, 190
novelty, 103, 151
number, 153
nutrient, 40, 172, 174, 175, 177

Index

nutrition, 38, 171, 173–178

oasis, 69
obesity, 16
object, 20
ocean, 145
offering, 2, 11, 44, 60, 71, 90, 97, 100, 103, 104, 119, 125, 159, 164, 165, 175, 190
office, 4, 69
Oldenburg, 89
one, 27, 56, 80, 81, 83, 87, 100, 106, 108, 110, 112, 117, 118, 158, 160, 179
operation, 153
operator, 58
opinion, 59
opportunity, 8, 93, 94, 162, 184
option, 29, 41, 125, 174
order, 22, 142
organization, 22, 29, 181, 182
origin, 15, 87, 142
other, 5, 16, 21, 23, 24, 28, 38, 39, 42, 44, 46, 54–56, 65, 68, 69, 77, 78, 85, 86, 98, 101, 103, 109, 110, 118, 123, 127, 144, 157, 158, 162, 163, 165, 172, 175–177, 181, 185, 186
outcome, 168
outdoor, 35, 36, 86–88, 102, 163, 164
outlet, 105, 112, 113
output, 55, 150
oven, 85
overall, 6, 7, 9, 21, 29, 38, 44, 71, 78, 111, 112, 116–119, 128, 145, 147, 150, 157, 159, 160, 162–164, 166, 169, 173, 175, 176, 178, 183, 185, 186
overindulging, 39
overview, 18, 22
owner, 98
oxytocin, 100

pace, 183
package, 122
packaging, 6–8, 12–15, 45, 92, 93, 95, 104, 122–124, 126–128, 144–146, 150, 152
pad, 162, 163
pain, 159, 160
pair, 20, 65, 70
palatability, 175
palate, 5, 167
pan, 80
panning, 61
paper, 12, 45
paradigm, 150
part, 3, 32, 34, 39, 40, 52, 54, 55, 82, 84, 86, 103, 161, 168, 177
participant, 90
participation, 94, 175, 184, 186
partner, 94
partnership, 189
passage, 89
past, 100
pastry, 94
pathway, 187
patient, 52
peace, 107, 119
peanut, 87
people, 1, 3, 4, 31, 32, 36–38, 51, 58–62, 86, 87, 91, 94, 101, 105, 107–110, 185, 188, 189, 191, 192

pepper, 80
peppermint, 36, 86
perception, 29, 52, 166, 185, 190
perfection, 85
performance, 61, 70, 71, 89, 104, 153, 157, 159
period, 29, 30, 153
persistence, 98, 189
perspective, 24, 84
perspiration, 161
pest, 48
pesticide, 12
phase, 68
phenomenon, 171
phrase, 87
physiology, 71, 159
picking, 153
pie, 85
piece, 53, 89, 104
pig, 121
pinna, 160
pity, 191
pivot, 161
place, 3, 34–36, 54, 78, 90, 99, 101, 104, 153, 158
placement, 166
placing, 153, 163
planar, 158
planet, 142, 143
plant, 1, 2, 12, 13, 23, 25, 27–29, 41, 43, 45, 77–80, 103, 123–127, 130, 144, 156
plantation, 124
plastic, 12, 15, 30, 138, 139, 144
plate, 158
platform, 59, 104, 105
play, 3, 4, 7, 8, 34–36, 55, 62, 83, 85, 92, 93, 97–99, 102, 106, 116, 121, 141, 145, 146, 150, 154, 157, 159, 162, 166, 168, 171, 174, 177, 181–183, 186, 187
playback, 57, 60, 69, 72, 73, 164
playfulness, 42, 89, 168
playtime, 113
pleasure, 100, 159, 179
pliability, 33
point, 36, 69, 159
pollination, 11, 42
pollution, 12–16, 125, 139, 145
popsicle, 87
popularity, 2, 16, 26, 29, 36, 37, 41, 44, 45, 56, 60–62, 66, 71, 80, 92, 100, 102, 103, 111, 116, 126, 160, 174
population, 1, 26, 31
porridge, 176
portability, 32
portion, 39, 40
position, 61, 161
positioning, 92
possibility, 56
potato, 26, 80, 85
potential, 2, 6, 11–13, 18, 21, 26, 29, 34, 38, 46–48, 52, 58, 62, 71, 102, 106, 108–111, 114, 116, 124, 138, 141, 142, 145, 146, 152, 156, 162, 166–169, 172, 176, 177, 179, 185, 187
poverty, 171–173
powder, 77, 78, 83
power, 13, 14, 32, 45, 56, 58, 59, 69, 99–101, 145, 157, 163, 167
practice, 29, 98, 100, 101, 109, 110, 117, 118, 127, 183, 184
prayer, 83

precision, 44, 52, 125, 152, 157
preference, 42
premium, 37, 41, 61
preparation, 85, 172
presence, 19, 32, 34, 83, 92, 119
present, 4, 17, 33, 57, 100, 101, 117, 118, 174
presentation, 167
preservation, 60, 78, 79
pressure, 142, 160–162
pretzel, 97
prevalence, 176
price, 8, 36, 69, 71, 131, 159
pricing, 143
principle, 53, 68, 126, 165
printer, 26
printing, 26
privacy, 4, 106–108, 110
problem, 12, 34, 88, 98
process, 13, 20, 24, 26–28, 30, 43, 44, 46, 55, 62, 65, 77, 86, 90, 98, 99, 112, 121–125, 127, 136, 138, 139, 143, 144, 150–153, 163, 167, 168, 180, 182, 183
processing, 12, 60, 62, 111–113, 121, 138, 139, 156, 163, 187
processor, 68
procurement, 141
product, 6, 7, 9, 14, 92–95, 103, 121, 122, 133, 138, 140, 141, 151–154
production, 1, 2, 6–16, 18–21, 23–26, 28–32, 37, 38, 43–48, 52, 55, 60–65, 79, 99, 104, 121–125, 127, 128, 138, 139, 143–145, 148, 150–154, 157, 167, 175, 178, 179
productivity, 44, 54, 69, 107
professional, 4, 20, 28, 62, 63, 65, 101, 186
profile, 83, 179
profitability, 6, 48, 142
program, 144, 176, 178, 182
progress, 174
project, 178
prominence, 33, 104
promise, 167–169, 187
promotion, 174
pronunciation, 182–184
propaganda, 59, 60
proprioception, 112
protein, 125, 172, 175, 176
proximity, 110
psychologist, 109
public, 4, 31, 59, 94, 107, 109, 110, 181
pumpkin, 85
punk, 105
puppet, 98
purchase, 93, 136, 179, 181
purchasing, 8, 19, 92, 93, 134, 136–138, 140, 143
puree, 103
purification, 45, 78
purpose, 5, 171, 177
pursuit, 62

quality, 2, 7–9, 11, 16, 19, 21, 30, 37, 41, 43–45, 48, 54, 55, 57, 60–62, 65, 69, 71, 80, 89, 103, 121, 124, 145, 147, 150–154, 157–160, 163–167, 185–187
quantity, 28, 48

race, 189
radiator, 159
radio, 2, 18, 31, 56, 58–60, 163, 165
range, 11, 17, 36, 43, 54, 61, 70, 73, 95, 99, 112, 131, 136, 148, 149, 157, 159, 161–163, 167, 168, 187
raspberry, 43
rate, 24, 30, 150, 161
ratio, 61, 154
reach, 7, 16, 52, 58, 70, 93, 94, 123, 165
reactivity, 154
reader, 168
reading, 183
realism, 159
reality, 11, 57, 58, 89, 166
realm, 11, 33, 89, 167, 168
reassurance, 35
receiver, 53
receiving, 53
reception, 58, 166, 167
recipe, 35, 81, 83, 94, 168
recognition, 55, 56, 94, 184, 191
record, 32, 60
recording, 2, 18, 55, 60, 62–65
recyclability, 21, 140
recycling, 8, 12–14, 16, 19, 45, 123, 126, 139, 141, 144–146
reduction, 70, 108, 113, 116, 123, 124
refinement, 26, 30
refining, 28
reflection, 105, 111, 114, 116
reforestation, 130
refugee, 178
regime, 59
regimen, 168
region, 4, 35, 48, 83, 84, 160

regulation, 100, 112, 113
relationship, 24, 33, 106
relaxation, 100, 102, 106, 111–114, 116, 118, 168, 169
release, 16, 45, 100, 107, 112, 139
relevance, 179
reliability, 17
reliance, 26, 43, 124, 125, 139, 142, 173
relief, 169, 172, 178, 179
remedy, 3, 24, 29
repair, 7, 14
repairability, 14
replacement, 65
representation, 21, 35, 188–192
reproduction, 2, 7, 18, 51–55, 57, 60, 61, 70, 158
reputation, 7, 8, 133, 190
research, 7, 8, 14, 17, 22, 48, 78, 106, 116, 145, 146, 156, 169, 189
resilience, 33, 98
resistance, 105, 156, 161, 162
resolution, 61
resonance, 159, 166
resource, 20
respect, 77, 78, 107, 142
respite, 101
response, 53, 61, 65, 144, 158, 159, 165
responsibility, 19, 130, 169
restaurant, 98, 166
restraint, 83
result, 7, 14, 15, 28, 29, 32, 68, 83, 109
resurgence, 29, 103
retention, 180
retreat, 4, 106, 107, 110
retrieval, 159

Index 215

return, 145, 152
reusing, 16, 123
revenue, 5–8, 21, 127–130
revolution, 18, 61
reward, 175
rhythm, 183
rice, 80, 83
ride, 4
ring, 159
rise, 2, 4, 8, 11, 32, 37, 38, 41,
    57–59, 61, 103, 131
risk, 15, 150, 152, 153, 172, 175
roast, 35
roasting, 31–33, 35, 86–88, 100
robotic, 153
rock, 105
role, 3–9, 18, 19, 21, 29–32, 34–36,
    44, 51, 52, 55, 57, 59, 62,
    79, 83, 85, 86, 92, 93,
    97–99, 101, 102,
    104–106, 116, 117, 121,
    141, 145, 146, 150, 157,
    159, 162, 166, 171, 174,
    177, 179–183, 185–187
root, 24, 25, 41, 78
rosemary, 37
rosewater, 79, 83
royalty, 24, 25
rubber, 87, 138
rumble, 68
run, 87

sadness, 100
safety, 35, 70, 88, 109, 113, 145,
    152, 153, 155, 156, 165,
    166, 169
sampling, 93
San Juan, 85
sanctuary, 4, 107

sanitation, 113
sanitization, 152
sap, 1, 23, 25, 26, 79
satisfaction, 54, 91, 162
saving, 14, 139
scalability, 145, 167
scale, 1, 26, 28, 30, 89, 94, 124, 150,
    151, 154, 174, 178
scarcity, 177
scent, 112
science, 17, 42, 52, 54, 99, 146, 154
sculpture, 34, 89, 90, 94, 95, 102
seal, 70, 152
sealing, 70
season, 85, 86
seaweed, 43
section, 1, 3, 5, 9, 12, 15, 20, 23, 25,
    32, 34, 36, 38, 41, 43, 46,
    51, 53, 55, 58, 60, 62, 66,
    68, 71, 77, 79, 82, 84, 86,
    89, 91, 92, 97, 99, 101,
    104, 106, 108, 111, 114,
    116, 121, 124, 127, 131,
    134, 138, 141, 144, 146,
    147, 150, 154, 157, 159,
    162, 167, 171, 177, 179,
    182, 185, 188
sector, 6, 129, 179
security, 15, 35, 110, 171, 173, 174,
    177, 179
segment, 131
segmentation, 6
selection, 44
self, 4, 5, 19, 101, 104–107, 111,
    113, 114, 116, 184
sender, 53
sensation, 36, 97, 112
sense, 4, 16, 33, 35, 60, 61, 86, 89,
    92, 93, 99, 100, 102, 103,

105–107, 109, 110, 116, 117, 119, 188
sensitivity, 78, 189
sensory, 5, 11, 34, 36, 42, 44, 97, 99, 100, 104, 111–113, 118, 119, 155, 156, 167–169, 180
separation, 61
series, 122, 123, 136, 151
serotonin, 100
service, 14, 54
set, 24, 26, 52, 80, 116, 122
setting, 63
setup, 53, 62, 65
shape, 5, 20, 21, 33, 42, 59, 62, 70, 82, 89–91, 98, 105, 133, 134, 138, 151, 152, 159, 160
shaping, 13, 24, 30–32, 44, 58, 59, 79, 92, 104, 112, 122–124, 166
share, 15, 35, 59, 94, 111, 132, 151, 164, 190
sharing, 35, 37, 94, 102, 105, 109, 111, 142
Shauna Sever, 84
shelf, 12, 122, 155, 156, 172, 174, 177
shift, 2, 31, 32, 49, 56, 61, 123, 150
ship, 58
shipping, 123
shooter, 87
shop, 69
shopping, 137
show, 32
side, 85
sight, 112
signal, 53, 54, 60–62, 68, 157, 158, 163, 187
signature, 122, 158, 159
significance, 1, 3, 5, 6, 9, 17–20, 25, 34, 36, 49, 58, 59, 77–79, 82, 84, 86, 89, 92, 99, 106, 124, 141, 179, 188
silence, 69
silica, 155
silver, 52, 155
simulation, 99
sinking, 58
situation, 171, 174
size, 19, 39, 89, 90, 131, 132, 151, 155, 160
sizing, 152
skewer, 86
skill, 99
skin, 121, 125, 160
smartphone, 163
smell, 112, 113
snack, 172
soaking, 43, 122
society, 1, 3, 5, 13, 21, 31, 58, 59, 92, 108, 141, 142, 191, 192
socio, 122
softness, 33, 35, 112
software, 73, 184
soil, 12, 13, 15, 124, 125, 139
solace, 32, 35, 100, 101, 107, 114, 193
solar, 13, 14, 45, 145
solitude, 4, 106–108
solution, 142, 153, 163, 187
solving, 34, 88, 98
song, 33, 104
sophistication, 95, 105
sound, 2, 5, 8, 11, 18, 19, 51–57, 60–63, 68–71, 74, 108, 112, 114–116, 119,

157–160, 164–169,
 185–187
soundscape, 118
soundstage, 159
soundtrack, 4, 5
source, 13, 56, 57, 90, 104, 124, 163,
 167, 172, 175–177, 179
sourcing, 7, 12–14, 19, 21, 121–124,
 127, 130, 139, 141–145
space, 4, 106, 107, 110
spaceship, 98
spacing, 153
spaciousness, 61
Spain, 85
spatiality, 60
speaker, 53, 184
speaking, 183, 184
specialty, 80
spectator, 90
spectatorship, 90
speech, 184–186
speed, 30, 44, 45, 122, 150, 153,
 157, 158
spending, 101
spirulina, 125
spoilage, 155
spread, 25, 59, 80, 113
spring, 27, 161
stability, 26, 99, 154–156, 163
stabilizer, 122
stage, 4, 52, 65, 110, 122, 151, 153
standardization, 30
staple, 3, 4, 102, 171, 173–175, 177,
 179
starch, 24–26, 44, 125
state, 48, 118
statement, 5
status, 61, 176
steam, 30

steeping, 44
step, 122, 142, 168
stereo, 2, 18, 54, 60–62
stethoscope, 52
stewardship, 141
stick, 86
sticking, 24
stone, 89
storage, 152, 156
story, 98, 193
storytelling, 59, 98, 99, 189
strain, 160
strategy, 92, 93, 127
strawberry, 36, 42, 43, 122
streaming, 61, 162, 163
strength, 105, 112
stress, 35, 100, 101, 107, 108, 112,
 113, 116, 117, 168, 169,
 183
structure, 20, 22, 26, 33, 98, 122,
 155, 156
student, 182
studio, 62, 65
study, 78, 106, 109, 126, 134, 140,
 152, 154, 156, 159, 181
style, 5, 105
sub, 32
subject, 17, 22
substance, 27, 28, 43
substitute, 43, 172
success, 6, 8, 45, 143, 145, 146, 178
sugar, 1, 6, 12, 16, 20, 23–26, 28, 37,
 38, 41, 42, 79, 83, 91, 103,
 121–124, 127, 128, 142,
 147, 151, 175, 177
sugarcane, 121, 124
suitability, 113, 172
summary, 90, 116, 123
summer, 102

sundae, 89
sunlight, 28
supplement, 175
supply, 2, 15, 16, 26, 93, 121, 139, 142, 178
support, 99, 101, 105, 113, 130, 143, 182, 188
surface, 154
surge, 37, 80
surgery, 52
surprise, 85, 104, 168, 171
surrounding, 16, 21, 26, 36, 69, 78, 92, 107, 110, 113, 160, 174
suspension, 155
sustainability, 2, 3, 7–9, 19, 21, 43, 45, 92, 104, 122–130, 143–146, 176, 179
sustenance, 172
sweat, 161, 162
sweating, 160, 162
sweaty, 161
sweetener, 122
sweetness, 35, 80, 82–86, 124
symbol, 1, 4, 18, 32–35, 61, 82, 86
symbolism, 90
symphony, 167
syrup, 6, 20, 24, 28, 38, 41, 80, 88, 121–125, 127, 128, 151, 177
system, 13, 24–26, 60, 68, 127, 153, 158, 175

t, 38, 87
table, 83, 189
tablet, 163
tactic, 93
tailor, 37, 138
talent, 189
talk, 109, 110

tangyuan, 83
tape, 33
tapestry, 5
tapioca, 44, 125
tapping, 95
target, 92, 95, 183, 184
task, 30, 142
taste, 3, 5, 28, 34, 36, 38, 40–42, 44, 45, 82, 83, 86, 90, 93, 99, 101, 103, 112, 113, 117, 118, 121, 124, 125, 147–151, 158, 175
Taylor Swift, 33
tea, 41, 44, 77, 83
teaching, 166, 182
teamwork, 33
tech, 57
technique, 25, 26, 43, 44, 60
technology, 2, 3, 5, 7, 10, 11, 16–19, 21, 29, 31, 42, 43, 52, 54, 55, 57, 60–62, 66, 68, 69, 71, 72, 74, 130, 131, 133, 142, 150, 154, 158, 159, 161–165, 167–169, 181, 184–187
teenager, 4
telephone, 2, 18, 31, 53–57
telephony, 52–55
television, 32, 34
temperature, 24, 44, 48, 122, 151, 152, 161, 162
tension, 87, 113
term, 7, 16, 48, 60, 71, 129, 130, 143, 152, 154, 159, 172, 174
terminology, 18
testament, 38
testing, 7
text, 22

Index

texture, 1, 3, 5, 19, 20, 24–26, 28, 29, 34, 36, 38, 40, 43–45, 80, 82, 83, 86, 89, 97, 100–102, 112, 113, 117, 118, 121, 122, 124, 125, 151, 152, 154–156, 171, 175
Thailand, 84
the Middle East, 84
the Middle Eastern, 83
the Soviet Union, 59
the United Kingdom, 79
the United States, 3, 48, 59, 80
theme, 168
themed, 94, 95, 102
theory, 17
therapy, 19, 111–114, 116, 168, 169
thickening, 78
thinking, 20, 98, 99, 111, 180
Thomas Edison, 54
thought, 78, 90, 190
thrive, 45, 124
throat, 27
tie, 93
tight, 70
time, 25–28, 35, 51, 54, 57, 85, 89, 90, 94, 101, 106, 111, 150, 151, 153, 162, 164, 168, 184
tinnitus, 16
tiramisu, 42
today, 25, 27–29, 31, 51, 56, 59, 108, 141, 150
togetherness, 4, 86, 102
tokenism, 189
tool, 54, 59, 60, 92, 95, 103, 105, 108, 112–118, 168, 172, 174, 179, 182, 184
toolkit, 101

top, 85
topic, 18, 65, 111, 190
topping, 1, 20, 80, 85, 93
touch, 80, 84–86, 89, 90, 95, 104, 112, 164
tower, 34, 98, 99
tracking, 2, 65
tract, 27, 166
traction, 92
trade, 7, 8, 12, 19, 21, 127, 130, 141–143, 179
tradition, 35, 85, 86, 92, 102
tragedy, 58
train, 4
training, 152, 153, 182
tranquility, 4
transfer, 153, 163
transformation, 29, 42, 90
transgender, 193
transience, 89
transmission, 52, 56–59, 165, 167, 187
transmitter, 166
transparency, 15, 70, 142, 143
transportation, 12, 107, 109, 123, 128, 139, 152
travel, 165
traveler, 70
treat, 1, 3, 24, 25, 29, 31, 34–36, 48, 78–84, 86, 87, 91, 99, 100, 102, 104, 171, 175
treatment, 27
trend, 37, 41, 80, 103, 104, 143, 163, 190
trendsetter, 95
trial, 98
trumpet, 51
trust, 100, 143, 151, 169
tube, 52, 56, 58

tune, 55
tunnel, 152
Turkey, 83
turn, 8, 32, 183
twist, 2, 4, 37, 83, 85, 149
type, 72, 83, 131, 157, 159

understanding, 17, 18, 20, 22, 40, 53, 59, 78, 79, 84, 98, 99, 106, 116, 117, 138, 179, 180, 183, 189, 190
uniformity, 155, 156
uniqueness, 5, 188
unity, 78
universe, 4
unrest, 59
upheaval, 35
usability, 52, 54, 63
usage, 4, 7, 11, 12, 19, 45, 65, 79, 106–111, 124–126
use, 3, 4, 8, 12, 13, 15, 16, 19, 21, 24, 26, 28, 29, 37, 43–45, 53–55, 57, 59, 66, 73, 77, 78, 80, 87, 90, 93, 95, 103, 106, 108, 110, 111, 114, 115, 118, 119, 122, 124, 140, 142, 145, 146, 149, 150, 154–156, 158, 160–163, 169, 177, 178, 180, 182–185, 187, 188
user, 21, 51, 53, 56, 67–70, 94, 131, 159–164, 185
utilization, 29, 169

vacuum, 58
value, 16, 21, 77, 99, 127, 130, 175, 177–179
vanilla, 36, 41, 43, 79, 80, 122

variety, 10, 27, 31, 33, 34, 36, 43, 45, 72, 85, 102, 147, 150
vegan, 6, 41, 43, 80, 130
vegetarian, 43, 80
vehicle, 5, 104
versatility, 25, 26, 29, 34, 35, 37, 68, 81, 82, 84, 86, 92, 93, 102, 104, 113
viability, 7, 48, 129, 130
vibration, 166
vice, 53
video, 184
Vietnam, 84
vinyl, 60
viscosity, 151
vision, 153
visual, 92, 93, 125, 180
vocabulary, 183
voice, 2, 54, 55, 57, 105, 157, 164
volume, 51, 56, 69, 70, 108, 154, 158, 164

wage, 141
war, 59
warehousing, 123
warmth, 4, 33, 35, 100
waste, 7, 8, 12–16, 21, 44, 45, 123, 124, 127, 138, 139, 141, 144–146, 153, 178
water, 7, 12, 13, 15, 24, 27, 28, 45, 48, 88, 121, 122, 124, 125, 139, 147, 165
wave, 68
way, 2, 4, 5, 11, 29, 31–33, 35, 41, 42, 52–56, 58–62, 66, 68, 74, 80, 84, 89, 93, 94, 100, 102, 108, 110, 112, 117, 146, 152, 164, 167, 189
weather, 2, 46, 48

Index

weight, 39, 99, 101, 160
welfare, 122, 125
well, 7, 15, 16, 18, 19, 22, 35, 38, 41, 45, 71, 78, 79, 85, 93, 99–101, 106–108, 111, 112, 114–117, 119, 125, 141–144, 152, 156, 158, 176, 178
wellness, 41, 91, 168
whimsy, 32, 34, 104
whipping, 20, 26, 30, 79, 122–124
white, 85, 109
wind, 13
winter, 35, 85
wire, 53, 157, 162
wireless, 7, 14, 18–21, 54, 57, 58, 62, 66–68, 72, 73, 131, 145, 162–165, 186, 187
wonder, 89
wood, 51
word, 79

work, 15, 68, 69, 79, 89, 90, 107, 113, 143
worker, 30
workforce, 153
working, 15, 65, 98, 142, 143, 181, 189
world, 2, 4, 5, 10, 17–20, 33, 34, 42, 43, 54, 55, 57, 58, 62, 66, 68, 78, 82, 84, 89–91, 97, 99, 102–104, 106–108, 116, 118, 126, 138, 141, 142, 146, 149, 162, 165, 167, 177, 179–181, 184, 185, 187
worldview, 105
wrapper, 144, 146
writing, 183

year, 93
Yoshinori Mizutani, 89
youth, 34
yuzu, 42

Milton Keynes UK
Ingram Content Group UK Ltd.
UKHW031617231124
451036UK00003B/37